Springer Proceedings in Physics

Volume 425

Indexed by Scopus

The series Springer Proceedings in Physics, founded in 1984, is devoted to timely reports of state-of-the-art developments in physics and related sciences. Typically based on material presented at conferences, workshops and similar scientific meetings, volumes published in this series will constitute a comprehensive up to date source of reference on a field or subfield of relevance in contemporary physics. Proposals must include the following:

– Name, place and date of the scientific meeting

– A link to the committees (local organization, international advisors etc.)

– Scientific description of the meeting

– List of invited/plenary speakers

– An estimate of the planned proceedings book parameters (number of pages/articles, requested number of bulk copies, submission deadline).

Please contact:

For Americas and Europe: Dr. Zachary Evenson; zachary.evenson@springer.com

For Asia, Australia and New Zealand: Dr. Loyola DSilva; loyola.dsilva@springer.com

Yahya Tayalati • Mohamed Gouighri
Editors

The First African Conference on High Energy Physics

Proceedings of ACHEP, Rabat & Kenitra, Morocco, October 23-27, 2023

 Springer

Editors
Yahya Tayalati
Faculty of Sciences
Mohammed V University in Rabat
Rabat, Morocco

Mohamed Gouighri
Department of Physics
Université Ibn-Tofail
Kénitra, Morocco

ISSN 0930-8989   ISSN 1867-4941   (electronic)
Springer Proceedings in Physics
ISBN 978-3-031-88932-5   ISBN 978-3-031-88933-2   (eBook)
https://doi.org/10.1007/978-3-031-88933-2

This work was supported by Sponsoring Consortium for Open Access Publishing in Particle Physics.

© The Editor(s) (if applicable) and The Author(s) 2026. This book is an open access publication.
Open Access This book is licensed under the terms of the Creative Commons Attribution 4.0 International License (http://creativecommons.org/licenses/by/4.0/), which permits use, sharing, adaptation, distribution and reproduction in any medium or format, as long as you give appropriate credit to the original author(s) and the source, provide a link to the Creative Commons license and indicate if changes were made.
The images or other third party material in this book are included in the book's Creative Commons license, unless indicated otherwise in a credit line to the material. If material is not included in the book's Creative Commons license and your intended use is not permitted by statutory regulation or exceeds the permitted use, you will need to obtain permission directly from the copyright holder.
The use of general descriptive names, registered names, trademarks, service marks, etc. in this publication does not imply, even in the absence of a specific statement, that such names are exempt from the relevant protective laws and regulations and therefore free for general use.
The publisher, the authors and the editors are safe to assume that the advice and information in this book are believed to be true and accurate at the date of publication. Neither the publisher nor the authors or the editors give a warranty, expressed or implied, with respect to the material contained herein or for any errors or omissions that may have been made. The publisher remains neutral with regard to jurisdictional claims in published maps and institutional affiliations.

This Springer imprint is published by the registered company Springer Nature Switzerland AG
The registered company address is: Gewerbestrasse 11, 6330 Cham, Switzerland

If disposing of this product, please recycle the paper.

Preface

The first African Conference on High Energy Physics, ACHEP2023, brings together scientists, researchers, and students from all over Africa and beyond to share the latest developments in high-energy physics. Organized by Ibn Tofail University and Mohammed V University, this conference represented a landmark for the African scientific community: a chance to collaborate, share knowledge, and celebrate Africa's fast-growing contribution to high-energy physics research worldwide.

The ACHEP2023 engaged its audience over several days with a rich program covering topics such as neutrino physics and dark matter searches, physics beyond the Standard Model, quark and lepton flavor physics, strong interactions and hadron physics, astrophysics and cosmology, and more. Sessions also explored detectors and future experimental facilities, applications of electroweak and Higgs physics, and heavy ions, as well as education and outreach in Africa. Highlights included updates on major international experiments like Hyper-Kamiokande, DUNE, and ATLAS. The conference featured over 95 oral presentations, 19 poster sessions, and keynote addresses delivered by leading international experts in their respective fields, offering insights into both current research and the future of high-energy physics.

Among the main ideas behind the ACHEP2023 was thus to give motivation and visibility to the next generation of physicists in Africa. Strong emphasis at the conference is also laid on the scientific infrastructure built across the continent, in particular, as a necessary instrument for their better integration into international collaboration. Separate sessions dedicated to issues of capacity building discussed policy decisions about further increasing the participation of African nations in large worldwide experiments by creating long-lasting possibilities for African scientists.

A cornerstone of ACHEP was its dedication to promoting gender balance and diversity. By prioritizing inclusivity, the conference created a platform that empowered underrepresented groups in physics, celebrated diverse perspectives, and inspired the next generation of scientists.

The Organizing Committee of ACHEP2023 warmly thanks all the participants, speakers, and contributors, for making this event a tremendous success. We would like to extend special thanks to our sponsors and host institutions whose support and vision have been indispensable in shaping this conference.

In publishing the proceedings of ACHEP2023, we celebrate the state-of-the-art research that was presented at this conference and look forward with bright hope toward the promising future of high-energy physics in Africa. We hope this volume will be an inspiration for further creativity, collaboration, and exploration.

Rabat, Morocco
Kénitra, Morocco

Yahya Tayalati
Mohamed Gouighri

Acknowledgments We are deeply honored to have received the esteemed patronage of His Majesty King Mohammed VI for The First edition of the African Conference of High Energy Physics. This prestigious recognition underscores His Majesty's enduring commitment to advancing science and technology, serving as a powerful testament to the importance of fostering innovation and international collaboration.

We acknowledge the kind support from the University Mohammed V in Rabat and the University Ibn Tofail in Kénitra, whose collaboration has been essential in ensuring the success of this conference. Furthermore, we extend our heartfelt thanks to the Centre National pour la Recherche Scientifique et Technique for their crucial support in organizing this event. Their dedication to advancing research and facilitating collaboration has played a key role in bringing this initiative to fruition.

Additionally, we are deeply grateful to the companies BERTIN and SUETS for their invaluable contributions. Their support has significantly enhanced the quality and scope of this conference, and we are honored to have their partnership in advancing our shared goals.

We also wish to express our sincere gratitude to the International Union of Pure and Applied Physics (IUPAP) for their endorsement of this event. Their support underscores the global significance of ACHEP and reinforces its role as a platform for fostering impactful scientific exchange and collaboration.

Contents

Part I Talks

Neutrino Spin States in Moving Matter and the Effect of Neutrino Spin Light... 3
Alexander Grigoriev, Alexander Studenikin, and Alexei Ternov

Wave Packet Treatment of Neutrino Flavour and Spin Oscillations in Galactic and Extragalactic Magnetic Fields................ 7
Artem Popov and Alexander Studenikin

LIGO–Virgo–KAGRA Results and Status of the Current Fourth Observing Run 13
David Keitel

Multi-Detector Analyses for CCSN Neutrino Detection 21
Meriem Bendahman, Isabel Goos, Joao Coelho, Matteo Bugli, Alexis Coleiro, Sonia El Hedri, Thierry Foglizzo, Davide Franco, Jérôme Guilet, Antoine Kouchner, Raphaël Raynaud, and Yahya Tayalati

Towards a Finite AdS_3 Topological Gravity Landscape 29
Rajae Sammani, Youssra Boujakhrout, El Hassan Saidi, Rachid Ahl Laamara, and Lalla Btissam Drissi

Maximinext Surfaces: Enhancements in Island Entropy 35
Safae Tariq, R. Ahl Laamara, and E. H. Saidi

A New Vertex Locator for the Next Generation LHCb Experiment 41
Enoch Ejopu

The Compact Linear Collider: Physics Potential 45
Jan Klamka

The Upgrade of the T2K Near Detector 49
Nataliya Skrobova

Hyper-Kamiokande: Physics Potential, Calibration and Detector Systematics.. 55
Sam J. Jenkins

Highlights of Recent Standard Model Measurements with ATLAS........ 61
Daniel Lewis

Search for Dark Matter with 2HDM+a in pp Collisions at $\sqrt{s} =$ 13 TeV with the ATLAS Detector 67
Sanae Ezzarqtouni

Jets Physics Simulations and Analysis with the ATLAS Experiment 75
Saad El Farkh

Reconstructing Long-Lived Particles with the ILD Detector 81
Jan Klamka

The Cross Section of Inverse Beta Decay 87
Giulia Ricciardi, N. Vignaroli, and F. Vissani

The KM3NeT Underwater Neutrino Telescope.............................. 93
Immacolata Carmen Rea

SATURNE: The Sarov Tritium Neutrino Experiment for Probing Coherent Elastic Neutrino-Atom Scattering and Neutrino Electromagnetic Interactions 99
M. Cadeddu, F. Dordei, C. Giunti, A. P. Ivashkin, K. A. Kouzakov, F. M. Lazarev, O. A. Moskalev, I. S. Stepantsov, A. I. Studenikin, I. I. Tkachev, V. N. Trofimov, M. A. Verkhovtsev, M. M. Vyalkov, A. A. Yukhimchuk, and E. F. Zagirdinova

An Overview of the DUNE Far Detectors 105
Leïla Haegel

Exploring Entanglement and Quantum Fisher Information in a System of Two Superconducting Qubits Subjected to Thermal Noise 111
Mourad Benzahra and Mostafa Mansour

Status and Perspective of ICARUS at the Fermilab Short-Baseline Neutrino Program... 117
Valerio Pia

Revisiting Inert Doublet Model Parameters 123
Hamza Abouabid, Abdesslam Arhrib, Ayoub Hmissou, and Larbi Rahili

Conformity of New CDF-II M_W with 123-Model........................... 129
B. Ait Ouazghour, R. Benbrik, Es-said Ghourmin, M. Ouchemhou, and L. Rahili

Enhanced Photon Higgs Associated Production in Two Higgs Doublet Type-II Seesaw Model at e^-e^+ Colliders 137
B. Ait Ouazghour, M. Chabab, and Khalid Goure

Collider Constraints on Massive Gravitons 147
Malak Ait Tamlihat, David d'Enterria, Laurent Schoeffel, Hua-Sheng Shao, and Yahya Tayalati

Results From the NA62 Experiment at CERN............................. 151
Mattia Soldani

Probing a Vector-like Top Quark $T \to H^+ b$ and $H^+ \to t\bar{b}$ at HL-LHC... 159
R. Benbrik, Mbark Berrouj, and Mohammed Boukidi

Interpreting the Indications of a 95 GeV Higgs Boson Through a 2-Higgs Doublet Model 165
A. Belyaev, R. Benbrik, Mohammed Boukidi, M. Chakraborti, S. Moretti, and S. Semlali

Search for New Phenomena with the ATLAS Detector at the LHC 173
Sijing Zhang

Recent Results from Belle and Belle II Experiments......................... 183
Paolo Branchini

Measurements of Heavy-flavour Production in pp Collisions with the ALICE Detector... 189
Tebogo Joyce Shaba

Part II Posters

The W Boson Mass Anomaly within the Inverted Scenario of the Two-Higgs Doublet Model.. 199
Hamza Abouabid, Abdesslam Arhrib, Rachid Benbrik, Mohamed Krab, and Mohamed Ouchemhou

Propagation and Oscillations of Cosmic Neutrinos in a Stochastic Magnetic Field.. 203
Konstantin Kouzakov, Anastasia Nikolaeva, and Alexander Studenikin

Weak/Strong Duality and the Asymptotic Weak Gravity Conjecture...... 207
Mohammed Charkaoui, R. Sammani, E. H. Saidi, and R. Ahl Laamara

Neutrino Oscillations and Quantum Decoherence 211
Konstantin Stankevich, Alexander Studenikin, and Maxim Vyalkov

Neutrino Oscillations in External Environment 215
A. Popov, N. Dolganov, V. Shakhov, Konstantin Stankevich, and A. Studenikin

Effect of Quantum Decoherence on Collective Neutrino Oscillations 219
Artem Popov, Anastasiia Purtova, Konstantin Stankevich,
and Alexander Studenikin

Hyper Kamiokande Energy Calibration with N16 223
Abderrazaq EL Abassi, Rafik Er-Rabit, and Mohamed Gouighri

Electromagnetic Effects in Deep Inelastic Neutrino-Proton Scattering 227
Konstantin Kouzakov, Elena Kovalevskaia, and Alexander Studenikin

**Dynamics of Quantum Resources in Hybrid System Under
Decoherence Effect** 231
Essalha Chaouki and Mostafa Mansour

**Elasic Neutrino-Nucleon Scattering: The Effects of
Electromagnetic Properties and Polarization** 237
Konstantin Kouzakov, Fedor Lazarev, and Alexander Studenikin

**Logarithmic Negativity Versus Quantum Discord in a System of
Dipolar Coupled Spins Undergoing Intrinsic Decoherence** 241
Mansoura Oumennana, Essalha Chaouki, and Mostafa Mansour

**Coherence and Non-classical Correlations within a Graphene
Layer System Subjected to Intrinsic Decoherence** 247
Zakaria Bouafia and Mostafa Mansour

Editors and Contributors

Editors

Mohamed Gouighri Professor Mohamed Gouighri is a distinguished physicist at the Department of Physics, University Ibn Tofail in Kenitra, Morocco. With extensive expertise in Elementary Particle Physics, Accelerator Physics, and Computational Physics, Dr. Gouighri has made substantial contributions to the global physics community.

Dr. Gouighri began his journey with The European Organization for Nuclear Research (CERN) as a Ph.D. student in 2007 and has maintained an active role in groundbreaking research ever since. His work spans B-physics, Higgs physics, and, more recently, neutrino physics. As a representative of Morocco in the Hyper-Kamiokande Collaboration, he serves as a member of their Steering and Resource Board while contributing to key projects such as Deuterium-Tritium Neutron Generator (DTG) calibration and physics analyses like Supernova sensitivity studies.

At University Ibn Tofail, Dr. Gouighri leads the university's A Toroidal LHC ApparatuS (ATLAS) group, fostering collaboration within one of CERN's flagship experiments. His research and leadership continue to inspire advancements in experimental physics, both hardware and software, while promoting international partnerships and the involvement of Moroccan physicists in global scientific endeavors.

Yahya Tayalati Professor Yahya Tayalati is a high-energy physicist, with over two decades of involvement in the ATLAS collaboration, the largest particle detection experiment at the CERN Large Hadron Collider. His contributions span from detector construction to the exploitation of physics data. He played a key role in the construction of the Liquid Argon Presampler and contributed to ground-breaking research on light-by-light scattering.

By building collaborations with the Astronomy with a Neutrino Telescope and Abyss environmental RESearch (ANTARES) and KM3NeT (KM Cube) neutrino telescope and the ATLAS High Granularity Timing Detector, Tayalati showcased

Morocco's role in cutting-edge particle physics. Recognized as an honorary fellow of the Islamic World Academy of Sciences, fellow of the World Academy of Sciences (TWAS) and a fellow of the African Academy of Sciences. In 2021, he was honored with the prestigious Mustafa Prize for his significant contribution to physics research and scientific development. With his leadership and expertise, Tayalati continues to shape the landscape of particle physics research, driving toward a deeper understanding of the fundamental principles of the Universe.

Contributors

Hamza Abouabid Faculty of Sciences and Techniques, Abdelmalek Essaadi University, Tangier, Morocco

Meriem Bendahman Université Paris Cité, CNRS, Astroparticule et Cosmologie, Paris, France

Faculty of Sciences, Mohammed V University in Rabat, Rabat, Morocco

Mourad Benzahra LHEPCM, Depart of Physics, Faculty of Sciences Ain Chock, Hassan II University, Casablanca, Morocco

Mbark Berrouj Polydisciplinary Faculty, Laboratory of Fundamental and Applied Physics, Cadi Ayyad University, Safi, Morocco

Zakaria Bouafia LHEPCM, Physics Department, Faculty of Sciences of Aïn Chock, Hassan II University, Casablanca, Morocco

Mohammed Boukidi Polydisciplinary Faculty, Laboratory of Fundamental and Applied Physics, Cadi Ayyad University, Safi, Morocco

Paolo Branchini RomaTre INFN Division, Rome, Italy

Essalha Chaouki LHEPCM, Depart of Physics, Faculty of Sciences Ain Chock, Hassan II University, Casablanca, Morocco

Mohammed Charkaoui LPHE-MS, Science Faculty, Mohammed V University in Rabat, Rabat, Morocco

Center of Physics and Mathematics, CPM, Rabat, Morocco

Enoch Ejopu The University of Manchester, Manchester, UK

Abderrazaq El Abassi Faculty of Sciences, Ibn Tofail University, Kenitra, Morocco

Saad El Farkh Faculty of Sciences, University Ibn Tofail, Kenitra, Morocco

Rafik Er-Rabit Faculty of Sciences, Ibn Tofail University, Kenitra, Morocco

Sanae Ezzarqtouni Faculty of Sciences Ain Chock, Hassan II University of Casablanca, Casablanca, Morocco

Es-said Ghourmin Laboratory of Theoretical and High Energy Physics, Faculty of Science, Ibn Zohr University, Agadir, Morocco

Khalid Goure LPHEA, Faculty of Science Semlalia, Cadi Ayyad University, Marrakech, Morocco

Alexander Grigoriev Moscow Institute for Physics and Technology, Dolgoprudny, Russia

Leïla Haegel Universite Claude Bernard Lyon 1, CNRS/IN2P3, IP2I Lyon, UMR 5822, Villeurbanne, France

Ayoub Hmissou Faculty of Sciences and Techniques, Abdelmalek Essaadi University, Tangier, Morocco

Sam J. Jenkins University of Liverpool, Liverpool, UK

David Keitel Departament de Física, Universitat de les Illes Balears, IAC3–IEEC, Palma, Spain

Jan Klamka Faculty of Physics, University of Warsaw, Warsaw, Poland

Konstantin Kouzakov Faculty of Physics, Lomonosov Moscow State University, Moscow, Russia

Elena Kovalevskaia Faculty of Physics, Lomonosov Moscow State University, Moscow, Russia

Daniel Lewis LAPP, Annecy, France

Fedor Lazarev Lomonosov Moscow State University, Moscow, Russia

Anastasia Nikolaeva Faculty of Physics, Lomonosov Moscow State University, Moscow, Russia

Mansoura Oumennana LHEPCM, Depart of Physics, Faculty of Sciences Ain Chock, Hassan II University, Casablanca, Morocco

Valerio Pia INFN Sezione di Bologna,, Bologna, Italy

Artem Popov Faculty of Physics, Lomonosov Moscow State University, Moscow, Russia

Immacolata Carmen Rea INFN Sez. di Napoli, Complesso universitario di Monte S. Angelo, Napoli, Italia

Giulia Ricciardi Dipartimento di Fisica E. Pancini, Università di Napoli Federico II and INFN, Naples, Italy

Rajae Sammani LPHE-MS, Science Faculty, Mohammed V University in Rabat, Rabat, Morocco

Centre of Physics and Mathematics, CPM, Rabat, Morocco

Tebogo Joyce Shaba North-West University, North-West, South Africa

iThemba LABS, Somerset West, Western Cape, South Africa

Nataliya Skrobova INR, Institute for Nuclear Research, Moscow, Russia

LPI, Lebedev Physics Institute, Moscow, Russia

Mattia Soldani INFN Laboratori Nazionali di Frascati, Frascati, Italy

Alexander Studenikin Moscow Institute for Physics and Technology, Dolgoprudny, Russia

Malak Ait Tamlihat Faculty of Sciences, Mohammed V University in Rabat, Rabat, Morocco

Safae Tariq LPHE-MS, Science Faculty, Mohammed V University in Rabat, Rabat, Morocco

Centre of Physics and Mathematics, CPM, Rabat, Morocco

Maxim Vyalkov Faculty of Physics, Lomonosov Moscow State University, Moscow, Russia

Branch of Lomonosov Moscow State University in Sarov, Sarov, Nizhny Novgorod Region, Russia

Sijing Zhang Southern Methodist University, Dallas, TX, USA

Laboratoire des 2 Infinis, Toulouse, France

Part I
Talks

Neutrino Spin States in Moving Matter and the Effect of Neutrino Spin Light

Alexander Grigoriev, Alexander Studenikin ⓘ, and Alexei Ternov

Neutrino electromagnetic properties and its interaction with matter are the key questions in physics of neutrino propagation in astrophysical media. The electromagnetic properties can be described by neutrino electromagnetic form factors [1] and the interaction with matter on the quantum level can be accounted for by the modified Dirac equation [2]. Combining these two research lines gives the so-called "method of exact solutions" [2, 3] which makes it possible to describe several new effects of neutrino propagation under the influence of external conditions (background matter and external electromagnetic fields). The method implies the use of the modified Dirac equation solutions to calculate amplitudes for particular processes.

The most important neutrino electromagnetic characteristic is the magnetic moment. In the minimally extended Standard Model with right-handed neutrinos its value is calculated to be $\mu \approx 3.2 \times 10^{-19} \mu_B (m/1 \text{ eV})$, where m is the neutrino mass and μ_B is the Bohr magneton [4]. Severe upper limits obtained in neutrino scattering experiments with the reactor and solar neutrino fluxes are at the level of $\mu \leq (2.8-2.9) \times 10^{-11} \mu_B$ [5, 6]. Quite recently significant progress has been made [7] in verifying the neutrino magnetic moments by analyzing data on the direct detection of dark matter in experiments.

In the present short note we aim to demonstrate the potential of the exact solution method which encompasses description of various phenomena related

---

Supported by the Scientific Program of the National Center for Physics and Mathematics, Section No. 8 (Stage 2023–2025).

A. Grigoriev (✉) · A. Ternov
Moscow Institute for Physics and Technology, Dolgoprudny, Russia

A. Studenikin
Lomonosov Moscow State University, Moscow, Russia
e-mail: studenik@srd.sinp.msu.ru

to the neutrino propagation in external media and the neutrino electromagnetic properties. On this basis the following important phenomena were predicted and studied.

1. The neutrino energy quantization in the case of neutrinos moving in a rotating matter [8].
2. When applied to millicharged neutrinos moving in a magnetized pulsar a new mechanism (*the neutrino star turning, νST*) that disturbs the star rotation angular velocity was predicted [9]. The comparison of the νST impact with the observed pulsars angular velocities provides the most stringent astrophysical bound on the neutrino millicharge $q_\nu < 1.3 \times 10^{-19} e_0$ (e_0 is the elementary charge).
3. The neutrino spin and spin-flavour oscillations can be engendered by neutrino interaction with the background moving matter in the case when there is the transversal matter current or matter polarization [10, 11].
4. A consistent treatment of neutrino oscillations in a magnetic field with account for massive neutrino helicity [12]. This approach reveals modulations of the neutrino oscillation amplitude owing to the interplay between vacuum and magnetic field oscillation frequencies.
5. The phenomenon of neutrino electromagnetic radiation in matter owing to its magnetic moment, called the Spin Light of Neutrino ($SL\nu$) [2, 13]. The process was studied for the case of non-moving matter and the higher the density of the matter and the neutrino energy, the more effective it is. Due to the latter fact it can be substantial in ultra-dense astrophysical environments (e.g., NSs, third-family compact stars, GRBs). For instance, for the case of third-family compact stars, assuming the baryon density as high as $n_b \sim 10^{41}$ cm^{-3} and neutrino energy $E \sim 10$ PeV, the radiation time is about $\tau_{SL\nu} \sim 10^{-4}$ s [14].

Here we show how the method of exact solutions can easily be adopted to asses the effect of transversal matter motion on $SL\nu$. This setting is peculiar for some GRB models. Note that the $SL\nu$ photon is emitted between different neutrino helicity states. In the limit of massless neutrinos they become the chiral states: ν_L for the initial neutrino and ν_R for the final one which is sterile. For the initial neutrino moving with momentum p and matter moving at a speed v along the third and first axes, respectively, we have the effective Dirac equation:

$$\{\gamma_0 E - \gamma_3 p - \tilde{n}(1 - \gamma_5)(\gamma_0 - v\gamma_1)\}\Psi_i(x) = 0, \qquad (1)$$

where matter is composed of electrons so that the density parameter is $\tilde{n} = \frac{1}{2\sqrt{2}} G_F(1 + 4\sin^2\theta_W)n$ with $n = \gamma n_0$ being the number density in the laboratory frame ($\gamma = 1/\sqrt{1-v^2}$ is the Lorentz gamma factor). For densities peculiar to the matter of neutron stars ($n \sim 10^{38}$ cm^{-3}) one has $\tilde{n} \sim 1$ eV, so that $\tilde{n}$ is very small compared to p. Then the neutrino energy leading from (1) can be written as $E = \sqrt{p^2 + 4\tilde{n}^2 v^2} + 2\tilde{n} \approx p + 2\tilde{n} + 2\tilde{n}^2 v^2/p$, and, for the standard representation of gamma-matrices, the solution to Eq. (1) has the form:

$$\Psi_i(x) = \frac{1}{\sqrt{2p}} (-\tilde{n}v, -p, \tilde{n}v, p)^T e^{-iEt+ipz}. \qquad (2)$$

For the final neutrino, the solution of the free Dirac equation is used.

Solving the kinematical problem with respect to the photon energy ω, while neglecting its dispersion for simplicity, gives: $\omega = \frac{2\tilde{n}p}{p(1-\cos\theta)+2\tilde{n}}$. The amplitude of the process is defined by the vertex $(\hat{\Gamma}e^*)$, where vector $\mathbf{e}$ describes the photon polarization and $\mathbf{\Gamma} = i\{[\mathbf{\Sigma} \times \mathbf{k}] + i\omega\gamma^5\mathbf{\Sigma}\}$ [2].

Within the straightforward calculation for the total rate and the power of $SL\nu$, in the limit $\tilde{n}/p \ll 1$ we obtain, respectively,

$$\Gamma = 4\mu^2\tilde{n}^2(p - 2\tilde{n} + \tilde{n}v^2) \quad I = \frac{4}{3}\mu^2\tilde{n}^2 p(p + 2\tilde{n}v^2). \qquad (3)$$

For the case of non-moving matter, $v = 0$, these formulas fit perfectly the results obtained before in [14]. Expressions (3) show that the $SL\nu$ in moving matter is γ^2 times more efficient than in the case of the non-moving one given that p and n_0 are the same. As a result, the $SL\nu$ radiation time is appeared to be γ^2 times less, which for high matter velocities in the real GRB conditions can give additional reduction by an order of magnitude.

Acknowledgments One of the authors (A.S.) is thankful to the organizers of the First African Conference on High Energy Physics for the hospitality. This study was supported by the Russian Science Foundation (project No. 22-22-00384).

References

1. C. Giunti, A. Studenikin, Neutrino electromagnetic interactions: A window to new physics. Rev. Mod. Phys. **87**, 531–591 (2015)
2. A. Studenikin, A. Ternov, Neutrino quantum states and spin light in matter. Phys. Lett. B **608**, 107–114 (2005)
3. A. Studenikin, Method of wave equations exact solutions in studies of neutrinos and electrons interaction in dense matter. J. Phys. A **41**, 164047 (2008)
4. K. Fujikawa, R.E. Shrock, Magnetic moment of a massive neutrino and neutrino-spin rotation. Phys. Rev. Lett. **45**, 963 (1980)
5. A.G. Beda et al., The results of search for the neutrino magnetic moment in GEMMA experiment. Adv. High Energy Phys. **2012**, 350150 (2012)
6. M. Agostini et al., Limiting neutrino magnetic moments with Borexino Phase-II solar neutrino data. Phys. Rev. D **96**, 091103(R) (2017)
7. C. Giunti, C.A. Ternes, Testing neutrino electromagnetic properties at current and future dark matter experiments. Phys. Rev. D **108**, 095044 (2023)
8. A. Grigoriev, A. Savochkin, A. Studenikin, Russ. Phys. J. **50**, 845 (2007)
9. A. Studenikin, I. Tokarev, Millicharged neutrino with anomalous magnetic moment in rotating magnetized matter. Nucl. Phys. B **884**, 396 (2014)
10. A. Studenikin, Neutrinos in electromagnetic fields and moving media. Phys. Atom. Nucl. **67**(993), 993 (2004)

11. P. Pustoshny, A. Studenikin, Neutrino spin and spin-flavour oscillations in transversal matter currents with standard and non-standard interactions. Phys. Rev. D **98**, 113009 (2018)
12. A. Popov, A. Studenikin, Neutrino eigenstates and flavour, spin and spin-flavour oscillations in a constant magnetic field. Eur. Phys. J. C **79**, 144 (2019)
13. A. Lobanov, A. Studenikin, Spin light of neutrino in matter and electromagnetic fields. Phys. Lett. B **564**, 27–34 (2003)
14. A. Grigoriev, A. Lokhov, A. Studenikin, A. Ternov, Spin light of neutrino in astrophysical environments. J. Cosmol. Astropart. Phys. **11**, 024 (2017)

Open Access This chapter is licensed under the terms of the Creative Commons Attribution 4.0 International License (http://creativecommons.org/licenses/by/4.0/), which permits use, sharing, adaptation, distribution and reproduction in any medium or format, as long as you give appropriate credit to the original author(s) and the source, provide a link to the Creative Commons license and indicate if changes were made.

The images or other third party material in this chapter are included in the chapter's Creative Commons license, unless indicated otherwise in a credit line to the material. If material is not included in the chapter's Creative Commons license and your intended use is not permitted by statutory regulation or exceeds the permitted use, you will need to obtain permission directly from the copyright holder.

Wave Packet Treatment of Neutrino Flavour and Spin Oscillations in Galactic and Extragalactic Magnetic Fields

Artem Popov and Alexander Studenikin

1 Introduction

It is known that massive neutrinos possess nontrivial electromagnetic properties, in particular nonzero anomalous magnetic moments. For a review of theory and experiment of neutrino electromagnetic properties see [1] and [2]. Currently, the best upper bound on neutrino effective magnetic moment obtained by the XENONnT experiment and is on the level $\mu_\nu \sim 6.4 \times 10^{-12} \mu_B$ [3, 4]. Interaction of neutrino magnetic moments with a magnetic field induces neutrino spin and spin-flavour precession, i.e. a phenomena in which neutrinos are converted to right-handed states. In particular, spin precession can be induced by neutrino interaction with cosmic magnetic field, that is of order of μG in our Galaxy [5].

To study neutrinos propagation through an environment of cosmological scales it is necessary to employ the wave packet approach that accounts for decoherence of neutrino oscillations. Previously, the wave packet formalism was developed for the case of vacuum neutrino oscillations within relativistic quantum mechanics (see [6] for a review) and quantum field theory [7] approaches, for neutrino oscillations in matter [8, 9] and collective neutrino oscillations [10]. In this paper we extend the wave packet approach to the case of neutrino propagation in a magnetic field.

A. Popov (✉) · A. Studenikin
Department of Theoretical Physics, Moscow State University, Moscow, Russia
e-mail: ar.popov@physics.msu.ru; studenik@srd.sinp.msu.ru

© The Author(s) 2026
Y. Tayalati, M. Gouighri (eds.), *The First African Conference on High Energy Physics*, Springer Proceedings in Physics 425,
https://doi.org/10.1007/978-3-031-88933-2_2

2 Neutrino Oscillations in a Magnetic Field in the Wave Packet Formalism

Neutrino oscillations in a uniform magnetic field are described by the following modified Dirac equation

$$(i\gamma^\mu \partial_\mu - m_i)v_i(x) - \mu_i \boldsymbol{\Sigma B} v_i(x) = 0, \quad (1)$$

where $i = 1, 2, 3$ enumerates neutrino massive states. In [11, 12], Eq. (1) was solved in the plane wave approximation. To account for potentially important decoherence effects in neutrino oscillations at long distances, we solve (1) assuming that at initial moment $t = 0$ neutrino wave function Fourier transform is described by a Gaussian wave packet

$$v_i(p, 0) = f_i(p, p_0) u_i^-(p), \quad f_i(p, p_0) = \frac{1}{(2\pi\sigma_p^2)^{1/4}} \exp\left(-\frac{(p - p_0)^2}{4\sigma_p^2}\right), \quad (2)$$

where p_0 is the average wave packet momentum, σ_p is wave packet width in momentum space and u_i^- is a left-handed solution of the Dirac equation in vacuum. For clarity, in this paper we consider one dimensional wave packets. Three-dimensional consideration leads to different wave packet spreading times in longitudinal and transversal directions, but does not significantly affect neutrino oscillations patterns [7, 13].

The dispersion relation for neutrinos interacting with a magnetic field is given by the following expression [11]

$$E_i^s(p) = \pm\sqrt{m_i^2 + p^2 + \mu_i^2 B^2 + 2s\mu_i \sqrt{m_i^2 B^2 + p^2 B_\perp^2}}, \quad (3)$$

where $s = \pm 1$. Here magnetic field $\boldsymbol{B}$ is decomposed into transverse $\boldsymbol{B}_\perp$ and longitudinal $\boldsymbol{B}_\parallel$ components with respect to the neutrino momentum.

For sufficiently small σ_p, (3) can be decomposed near average momentum p_0

$$E_i^s(p) = E(p_0) + v_i^s(p_0)(p - p_0) + \mathcal{O}((p - p_0)^2), \quad (4)$$

where $\left.\frac{\partial E_i^s(p)}{\partial p}\right|_{p=p_0}$ are group velocities. Using (3), we calculate neutrino wave packets group velocities accounting for neutrino interaction with a magnetic field

$$v_i^s(p_0) = \frac{p_0}{E_i^s(p_0)} \left(1 + \frac{s\mu_i B_\perp^2}{\sqrt{m_i^2 B^2 + p_0^2 B_\perp^2}}\right). \quad (5)$$

Finally, for the probabilities of neutrino flavour and spin oscillations in a magnetic field accounting for decoherence effects we get

$$P_{\nu_\alpha^h \to \nu_\beta^{h'}}(L) = \sum_{i,j} \sum_{s,s'} U_{\beta i}^* U_{\alpha i} U_{\beta j} U_{\alpha j}^* C_{is}^{hh'} C_{js'}^{hh'} \exp\left(-i2\pi \frac{L}{L_{osc}^{ijss'}}\right)$$

$$\times \exp\left(-\frac{L^2}{(L_{coh}^{ijss'})^2}\right), \tag{6}$$

where the corresponding oscillations and coherence lengths are introduced

$$L_{osc}^{ijss'} = \frac{\pi}{\omega_{ij}^{ss'}}, \quad L_{coh}^{ijss'} = \frac{2\sqrt{2}\sigma_x}{v_i^s - v_j^{s'}}. \tag{7}$$

The oscillations frequencies $\omega_{ij}^{s\sigma}$ are given by $\omega_{ij}^{s\sigma}(p_0) \approx \frac{\Delta m_{ij}^2}{2p_0} + (\mu_i s - \mu_j \sigma) B_\perp$. The coefficients $C_{is}^{hh'}$ in (6) are given by

$$C_{is}^{LL} \approx \frac{1}{2} + \mathcal{O}\left(\frac{m_i^2}{p^2}\right), \quad C_{is}^{RL} \approx -\frac{s}{2} + \mathcal{O}\left(\frac{m_i^2}{p^2}\right). \tag{8}$$

The oscillations probabilities (6) generalize the expressions we obtained in [12] and account for exponential damping of neutrino oscillations at large distance due to wave packets separation. Using (5), we obtain the following approximate expressions for the coherence lengths given that $p \gg m_i \gg \mu_i B$:

$$L_{coh}^{ijss} \approx \frac{4\sqrt{2}\sigma_x p^2}{\Delta m_{ij}^2}, \quad L_{coh}^{ii-+} \sim \frac{\sigma_x p^3}{\mu_i B m_i^2}, \quad L_{coh}^{ij-+} \approx L_{coh}^{ijss}. \tag{9}$$

Here the coherence lengths L_{coh}^{ijss} and L_{coh}^{ii-+} describe damping of oscillations on vacuum lengths $L_{osc}^{ijss} = \frac{4\pi p}{\Delta m_{ij}^2}$ and magnetic lengths $L_{osc}^{ii-+} = \frac{\pi}{\mu_i B_\perp}$, correspondingly. Note that unlike the coherence lengths of vacuum neutrino oscillations, the coherence length L_{coh}^{ii-+} is proportional to the cube of the neutrino momentum.

Interaction with a magnetic field can modify the flavour composition of neutrino emanating from astrophysical objects. For example, consider high-energy neutrinos originating from the Galactic centre. The flavour composition at distance L is given by

$$r_\alpha(L) = \frac{\sum_\alpha r_\alpha^0 P_{\alpha\beta}(L)}{\sum_{\alpha\beta} r_\alpha^0 P_{\alpha\beta}(L)}, \tag{10}$$

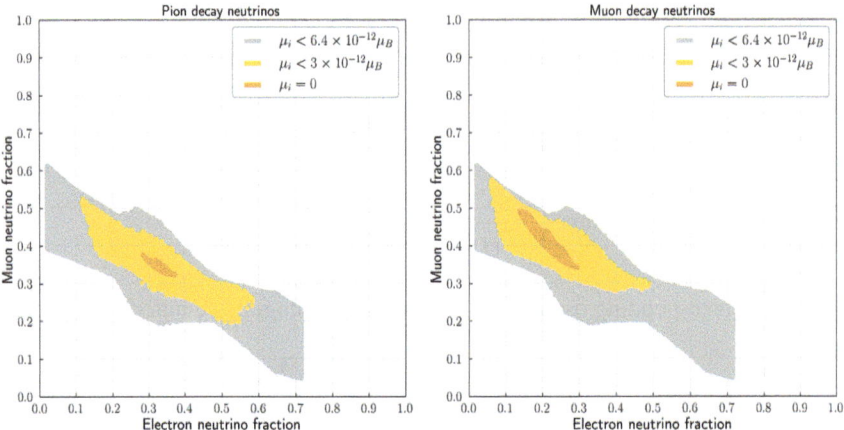

Fig. 1 Flavour compositions of neutrinos coming from the Galactic centre for the cases of pion decay and muon decay high-energy neutrinos production

where r_β^0 are the flavour ratios at the source and $P_{\alpha\beta} = P_{\nu_\alpha^L \to \nu_\beta^L}$ are the probabilities of flavour oscillations.

The mechanism of high-energy neutrinos production is presently unknown. We study two cases considered in the literature: case of pion decay neutrinos and case of muon decay neutrinos. The corresponding initial flavour compositions are $r^0 = (\frac{1}{3}, \frac{2}{3}, 0)$ and $r^0 = (0, 1, 0)$. We also consider different values of neutrino magnetic moments from the interval $(0.1, 6.4) \times 10^{-12} \mu_B$, as well as different experimental values of neutrino mixing parameters within their 3σ intervals. In Fig. 1 we show the possible flavour ratios of neutrinos propagating from the Galactic centre to a terrestrial neutrino telescope for different values of neutrino magnetic moments μ_1, μ_2, μ_3 and compare them to ones predicted by the vacuum neutrino oscillations case. We assume that galactic magnetic field strength is $B = 0.2 \ \mu G$, which corresponds to the average value of the regular component of the Galactic magnetic field. Magnetic field of approximately same strength is found in the Local Supercluster [14]. We find that the neutrino interaction with the Galactic magnetic field can significantly modify the observed flavour ratio. Thus, we conclude that neutrino magnetic moments potentially can be probed by neutrino telescopes, such as IceCube, Baikal-GVD and KM3NeT. The obtained results can be also applied for description of supernovae neutrino oscillations effects would be detected by JUNO and Hyper-Kamiokande.

This study was conducted within the Scientific Program of the National Center for Physics and Mathematics, Section No. 8 (Stage 2023-2025).

References

1. C. Giunti, A. Studenikin, Rev. Mod. Phys. **87**, 531 (2015)
2. A. Studenikin, PoS **CORFU2021**, 057 (2022)
3. E. Aprile et al. [XENON Collaboration], Phys. Rev. Lett. **129**(16), 161805 (2022)
4. C. Giunti, C.A. Ternes, Phys. Rev. D **108**(9), 095044 (2023)
5. R. Beck, AIP Conf. Proc. **1085**(1), 83–96 (2009)
6. C. Giunti, Found. Phys. Lett. **17**, 103–124 (2004)
7. D.V. Naumov, V.A. Naumov, Phys. Part. Nucl. **51**(1), 1–106 (2020)
8. J.T. Peltoniemi, V. Sipilainen, J. High Energy Phys. **06**, 011 (2000)
9. J. Kersten, A.Y. Smirnov, Eur. Phys. J. C **76**(6), 339 (2016)
10. E. Akhmedov, J. Kopp, M. Lindner, J. Cosmol. Astropart. Phys. **09**, 017 (2017)
11. A. Popov, A. Studenikin, Eur. Phys. J. C **79**(2), 144 (2019)
12. A. Lichkunov, A. Popov, A. Studenikin, [arXiv:2207.12285 [hep-ph]] (2019)
13. D.V. Naumov, Phys. Part. Nucl. Lett. **10**, 642–650 (2013)
14. J.P. Vallée, Astron. J. **124**, 1322–1327 (2002)

Open Access This chapter is licensed under the terms of the Creative Commons Attribution 4.0 International License (http://creativecommons.org/licenses/by/4.0/), which permits use, sharing, adaptation, distribution and reproduction in any medium or format, as long as you give appropriate credit to the original author(s) and the source, provide a link to the Creative Commons license and indicate if changes were made.

The images or other third party material in this chapter are included in the chapter's Creative Commons license, unless indicated otherwise in a credit line to the material. If material is not included in the chapter's Creative Commons license and your intended use is not permitted by statutory regulation or exceeds the permitted use, you will need to obtain permission directly from the copyright holder.

LIGO–Virgo–KAGRA Results and Status of the Current Fourth Observing Run

David Keitel

1 The LVK Detector Network and Collaboration

It took 100 years from Einstein's prediction of gravitational waves (GWs) to their first detection [3]. Any time-varying mass quadrupole can produce these ripples in spacetime. But we need both extreme astrophysical sources and extremely sensitive detectors to observe this new spectrum of cosmic messengers.

The current global network of laser-interferometric GW detectors consists of two LIGO detectors in the US with 4 km arms [1], the 3 km Virgo detector in Italy [25], the 3 km underground KAGRA in Japan [27] and the 0.6 km GEO600 in Germany [31], mainly used to test new technologies and watch for nearby GW events when the larger detectors are offline. Another 4 km LIGO will be constructed in India [37].

The detectors are operated and their data analyzed by the global LIGO–Virgo–KAGRA collaboration (LVK) with over 2000 members in over 200 groups. The progression of LVK observing runs and sensitivities is shown in Fig. 1.

2 GWs from Compact Binary Coalescences (CBCs)

The first detection of GWs [3] was the GW150914 signal from the merger of a binary black hole (BBH) with component masses $\sim 30\, M_\odot$ at a distance ~ 400 Mpc. Found by both matched-filter and unmodelled analysis pipelines, the observed

---

David Keitel for the LIGO Scientific Collaboration, the Virgo Collaboration, the KAGRA Collaboration.

D. Keitel (✉)
Departament de Física, Universitat de les Illes Balears, IAC3–IEEC, Palma, Spain
e-mail: david.keitel@ligo.org

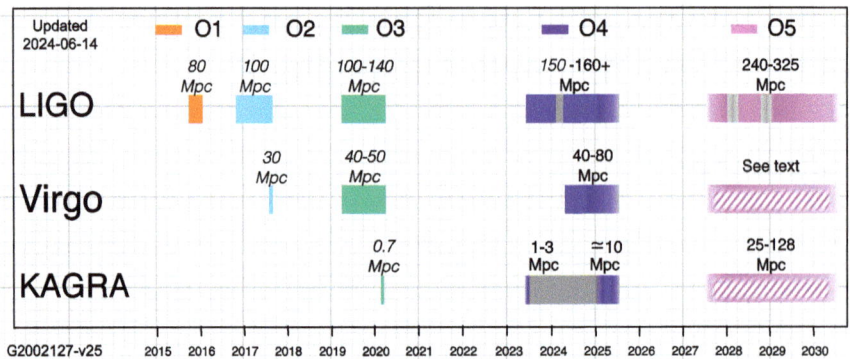

Fig. 1 LVK observing runs so far and plans for the rest of O4 and the future. Sensitivities are summarized as the typical distance at which binary neutron star mergers can be detected. Further updates will be available at https://dcc.ligo.org/G2002127/public/. © 2024, the LIGO Scientific Collaboration, the Virgo Collaboration and the KAGRA Collaboration

waveform closely matches full numerical relativity (NR) solutions of Einstein's equations [2].

Over the first three observing runs of Advanced LIGO, joined by Advanced Virgo since O2 and KAGRA at the end of O3, a total of 90 GW events have been published by the collaboration [9, 13, 20]. Additional events have been reported by external authors (e.g., [34, 40]) from open data provided by the LVK [17, 21]. All come from CBCs, and most from BBHs since heavier objects produce stronger GWs, and hence BBHs can be observed at larger distances. Our current data-based estimates of actual merger rates (per Gpc^3 and year) are 17.9–44 for BBHs, 10–1700 binary neutron stars (BNSs) and 7.8–140 for neutron star—black holes (NSBHs) [22]). The most distant observations reach redshifts of $z \sim 1$.

2.1 BBH Results

Since the GW frequency evolution depends on the source parameters (masses, spins, inclination, sky location), we can extract these parameters through Bayesian inference [39] against large numbers of NR-calibrated model waveforms (see references in [20]). We have found a wide spread of systems consistent with BBHs, recovering an increasingly sharp picture at the population level [22] of their rate and mass distribution. Observed component masses range from a few solar masses up into the intermediate-mass range ($\gtrsim 100\, M_\odot$), touching both of the "mass gaps" often discussed in astrophysics [22]. We have also found such systems with mass ratios from unity to 1:10 and with various spin configurations, though the latter are still difficult to measure precisely at current detector sensitivity. Both mass ratios and spins are useful indicators for astrophysical formation channels [22].

All events so far are well fit by general-relativity waveforms, but the LVK probes alternative theories of gravity through an array of tests yielding increasingly strong constraints [16]: inspiral-merger-ringdown consistency, parametrized tests of GW generation and GW dispersion relation, polarization tests, etc. We are also searching for signatures of gravitational lensing of GWs [15, 23].

2.2 Binaries Including Neutron Stars

The first BNS observation, GW170187 [6], was a revolutionary multimessenger event [7]. Among the many results were a first "standard siren" measurement of the Hubble constant [4, 19], confirmation that the speed of gravity is very close to that of light [5], constraints on the equation of state of nuclear matter at extreme densities [8, 9], and that BNS mergers are a prime source of heavy elements in the Universe [30]. A second BNS event, GW190425, hinted at a surprising population of heavy BNS systems [10, 20]. The set of CBC types observable with the LVK network was completed with the first observation of NSBHs [14, 20].

3 Beyond Binaries: Other GW Sources in the LVK Band

3.1 GW Bursts

While CBCs are well-modelled signals for which matched filtering is the standard approach, other sources of short GW transients ("bursts") are less well understood: core-collapse supernovae, magnetars, accretion disk instabilities, highly eccentric BBHs, cosmic strings, etc. Burst searches use more generic methods, often based on empirical models such as wavelets or sine-Gaussians or pattern recognition in time-frequency GW spectrograms. See [11, 12] for recent LVK all-sky burst searches, and references therein for physical scenarios and analysis methods. No detections of any signals beyond CBCs have been made so far, but even non-detections can yield interesting physical constraints, e.g., on galactic supernovae or on magnetars vs. BNS as sources for nearby gamma-ray bursts.

3.2 Continuous GWs

By contrast, long-duration persistent GWs can also be emitted by a number of sources, especially spinning non-axisymmetrically deformed neutron stars. They produce very weak GW strains but stay observable for years, with great promise for multi-messenger astronomy and as probes of nuclear physics at extreme densities. For a review of the physics, computationally very challenging search methods and the non-detection upper limits results obtained so far, see [36]. CW search methods are also used for direct and indirect dark matter detection with GW detectors; see Sect. 5.

3.3 Stochastic GW Backgrounds

GW detectors could uncover two types of stochastic backgrounds in the Universe: astrophysical backgrounds from the overlap of faint, unresolved CBCs and other sources or cosmological backgrounds from early-universe physics: inflationary tensor modes, phase transitions, etc. For recent reviews see [35, 38]. Complementary to LVK searches, in 2023 pulsar timing array experiments have found first evidence for such backgrounds at much lower (nHz) frequencies [26].

4 The Present: The O4 Run

The O4 run started on 27 May 2023, with sensitivity of the LIGO detectors further improved over O3. Virgo rejoined after a commissioning break between the O4a and O4b periods and KAGRA plans to participate again later in O4b, which will run into 2025. For detector status see gwosc.org/detector_status/O4a/.

Public alerts from low-latency data analysis are available via gracedb.ligo.org. News in O4 include alerts for marginal candidates (to enable deep multimessenger coincidence searches) and BNS pre-merger alerts. The accumulation of detections from O1–O3 and O4a candidates is shown in Fig. 2. When ACHEP2023 was held

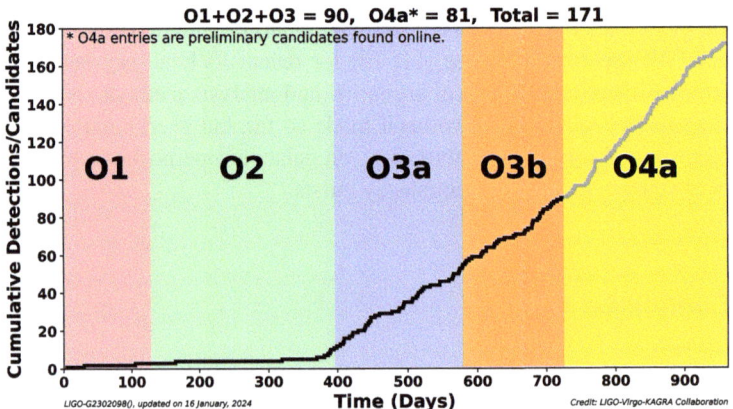

Fig. 2 GW detections published by the LVK from the O1–O3 runs [20, 24] and O4a candidates up to 2024-01-16, released to the public via gracedb.ligo.org. © 2024, the LIGO Scientific Collaboration, the Virgo Collaboration and the KAGRA Collaboration

in October 2023, there were 55 significant candidates from O4a; and by the time of the final version of this proceedings papers in June 2024, the number from O4a+b had grown to over 100. CBC catalog updates from O4 and further search results for bursts, CWs, stochastic backgrounds, etc. will be published by the LVK in due course, and O4 data will be made public via gwosc.org.

5 GWs, High-Energy Physics and New Physics

Of particular interest to the HEP community are the many avenues for exploring fundamental physics with GWs. These include the already mentioned tests of theories of gravity [16] and GWs as a new probe for cosmography [4, 19], the early Universe [33], and nuclear matter at extreme densities [8, 36]. LVK data can also be used to look for exotic compact objects [29], primordial BHs [32] and particulate dark matter—via indirect detection (GWs from annihilating boson clouds [28]) or direct detection of interactions with the detector hardware [18]. And multi-messenger astronomy [7] includes GWs and traditional photonic astronomy together with high-energy astroparticle observatories (gamma-ray, TeV and neutrino detectors). The HEP and GW communities also have notable overlap in data analysis techniques, computing infrastructure, instrumentation, the planning of large-scale facilities, as well as potential synergies in global community building and outreach strategies.

6 Conclusions

Through decades of work and with the expertise of a global community, GW astrophysics became reality. With 90 published detections from the first three observing runs and almost as many new candidates already from the ongoing O4 run, the LVK is obtaining unprecedented insights into the physics, populations and evolutionary history of compact objects in our Universe. Many other science targets are within reach as the global detector network is growing and continuing to improve its sensitivity, and future GW detectors will further push cosmic frontiers.

Acknowledgments D.K. was supported by the Universitat de les Illes Balears (UIB); the Spanish Agencia Estatal de Investigación grants CNS2022-135440, PID2022-138626NB-I00, RED2022-134204-E, RED2022-134411-T, funded by MICIU/AEI/10.13039/501100011033, the European Union NextGenerationEU/PRTR, and the ERDF/EU; and the Comunitat Autònoma de les Illes Balears through the Conselleria d'Educació i Universitats with funds from the European Union NextGenerationEU/PRTR-C17.I1 (SINCO2022/6719) European Union - European Regional Development Fund (ERDF) (SINCO2022/18146). This material is based upon work supported by NSF's LIGO Laboratory which is a major facility fully funded by the National Science Foundation. For LVK acknowledgements see dcc.ligo.org/P2100218/public. This paper has been assigned document number LIGO-P2300460-v3.

References

1. J. Aasi et al., Advanced LIGO. Class. Quantum Grav. **32**, 074001 (2015). https://doi.org/10.1088/0264-9381/32/7/074001
2. B.P. Abbott et al., Directly comparing GW150914 with numerical solutions of Einstein's equations for binary black hole coalescence. Phys. Rev. D **94**(6), 064035 (2016). https://doi.org/10.1103/PhysRevD.94.064035
3. B.P. Abbott et al., Observation of gravitational waves from a binary black hole merger. Phys. Rev. Lett. **116**, 061102 (2016). https://doi.org/10.1103/PhysRevLett.116.061102
4. B.P. Abbott et al., A gravitational-wave standard siren measurement of the Hubble constant. Nature **551**(7678), 85–88 (2017). https://doi.org/10.1038/nature24471
5. B.P. Abbott et al., Gravitational waves and gamma-rays from a binary neutron star merger: GW170817 and GRB 170817A. Astrophys. J. Lett. **848**(2), L13 (2017). https://doi.org/10.3847/2041-8213/aa920c
6. B.P. Abbott et al., GW170817: Observation of gravitational waves from a binary neutron star inspiral. Phys. Rev. Lett. **119**(16), 161101 (2017). https://doi.org/10.1103/PhysRevLett.119.161101
7. B.P. Abbott et al., Multi-messenger observations of a binary neutron star merger. Astrophys. J. Lett. **848**(2), L12 (2017). https://doi.org/10.3847/2041-8213/aa91c9
8. B.P. Abbott et al., GW170817: Measurements of neutron star radii and equation of state. Phys. Rev. Lett. **121**(16), 161101 (2018). https://doi.org/10.1103/PhysRevLett.121.161101
9. B.P. Abbott et al., GWTC-1: A gravitational-wave transient catalog of compact binary mergers observed by LIGO and virgo during the first and second observing runs. Phys. Rev. X **9**(3), 031040 (2019). https://doi.org/10.1103/PhysRevX.9.031040
10. B.P. Abbott et al., GW190425: observation of a compact binary coalescence with total mass $\sim 3.4 M_\odot$. Astrophys. J. Lett. **892**(1), L3 (2020). https://doi.org/10.3847/2041-8213/ab75f5
11. R. Abbott et al., All-sky search for long-duration gravitational-wave bursts in the third Advanced LIGO and Advanced Virgo run. Phys. Rev. D **104**(10), 102001 (2021). https://doi.org/10.1103/PhysRevD.104.102001
12. R. Abbott et al., All-sky search for short gravitational-wave bursts in the third Advanced LIGO and Advanced Virgo run. Phys. Rev. D **104**(12), 122004 (2021). https://doi.org/10.1103/PhysRevD.104.122004
13. R. Abbott et al., GWTC-2: Compact binary coalescences observed by LIGO and Virgo during the first half of the third observing run. Phys. Rev. X **11**, 021053 (2021). https://doi.org/10.1103/PhysRevX.11.021053
14. R. Abbott et al., Observation of gravitational waves from two neutron star-black hole coalescences. Astrophys. J. Lett. **915**(1), L5 (2021). https://doi.org/10.3847/2041-8213/ac082e
15. R. Abbott et al., Search for lensing signatures in the gravitational-wave observations from the first half of LIGO–Virgo's third observing run. Astrophys. J. **923**(1), 14 (2021). https://doi.org/10.3847/1538-4357/ac23db
16. R. Abbott et al., Tests of general relativity with GWTC-3 (2021 Dec), https://arxiv.org/abs/2112.06861
17. R. Abbott et al., Open data from the first and second observing runs of Advanced LIGO and Advanced Virgo. SoftwareX **13**, 100658 (2021). https://doi.org/10.1016/j.softx.2021.100658
18. R. Abbott et al., Constraints on dark photon dark matter using data from LIGO's and Virgo's third observing run. Phys. Rev. D **105**(6), 063030 (2022). https://doi.org/10.1103/PhysRevD.105.063030
19. R. Abbott et al., Constraints on the cosmic expansion history from GWTC–3. Astrophys. J. **949**(2), 76 (2023). https://doi.org/10.3847/1538-4357/ac74bb
20. R. Abbott et al., GWTC-3: Compact binary coalescences observed by LIGO and Virgo during the second part of the third observing run. Phys. Rev. X **13**(4), 041039 (2023). https://doi.org/10.1103/PhysRevX.13.041039

21. R. Abbott et al., Open data from the third observing run of LIGO, Virgo, KAGRA, and GEO. Astrophys. J. Suppl. **267**(2), 29 (2023). https://doi.org/10.3847/1538-4365/acdc9f
22. R. Abbott et al., Population of merging compact binaries inferred using gravitational waves through GWTC-3. Phys. Rev. X **13**(1), 011048 (2023). https://doi.org/10.1103/PhysRevX.13.011048
23. R. Abbott et al., Search for gravitational-lensing signatures in the full third observing run of the LIGO-Virgo network. Astrophys. J. **970**, 191 (2024). https://doi.org/10.3847/1538-4357/ad3e83
24. R. Abbott et al., GWTC-2.1: Deep extended catalog of compact binary coalescences observed by LIGO and Virgo during the first half of the third observing run. Phys. Rev. D **109**(2), 022001 (2024). https://doi.org/10.1103/PhysRevD.109.022001
25. F. Acernese et al., Advanced Virgo: a second-generation interferometric gravitational wave detector. Class. Quantum Grav. **32**(2), 024001 (2015). https://doi.org/10.1088/0264-9381/32/2/024001
26. G. Agazie et al., Comparing recent PTA results on the nanohertz stochastic gravitational wave background. Astrophys. J. **966**(1), 105 (2024). https://doi.org/10.3847/1538-4357/ad36be
27. T. Akutsu et al., Overview of KAGRA: Detector design and construction history. Prog. Theoret. Exp. Phys. **2021**(5), 05A101 (2021). https://doi.org/10.1093/ptep/ptaa125
28. R. Brito, V. Cardoso, P. Pani, Superradiance: new frontiers in black hole physics. Lect. Notes Phys. **906**, 1–237 (2015). https://doi.org/10.1007/978-3-319-19000-6
29. V. Cardoso, P. Pani, Testing the nature of dark compact objects: a status report. Living Rev. Rel. **22**(1), 4 (2019). https://doi.org/10.1007/s41114-019-0020-4
30. J.J. Cowan, C. Sneden, J.E. Lawler, A. Aprahamian, M. Wiescher, K. Langanke, G. Martínez-Pinedo, F.K. Thielemann, Origin of the heaviest elements: The rapid neutron-capture process. Rev. Mod. Phys. **93**(1), 15002 (2021). https://doi.org/10.1103/RevModPhys.93.015002
31. K.L. Dooley et al., GEO 600 and the GEO-HF upgrade program: successes and challenges. Class. Quant. Grav. **33**, 075009 (2016). https://doi.org/10.1088/0264-9381/33/7/075009
32. J. García-Bellido, Massive primordial black holes as dark matter and their detection with gravitational waves. J. Phys. Conf. Ser. **840**(1), 012032 (2017). https://doi.org/10.1088/1742-6596/840/1/012032
33. M. Maggiore, Gravitational wave experiments and early universe cosmology. Phys. Rep. **331**, 283–367 (2000). https://doi.org/10.1016/S0370-1573(99)00102-7
34. A.H. Nitz et al., 4-OGC: Catalog of gravitational waves from compact binary mergers. Astrophys. J. **946**(2), 59 (2023). https://doi.org/10.3847/1538-4357/aca591
35. A.I. Renzini, B. Goncharov, A.C. Jenkins, P.M. Meyers, Stochastic gravitational-wave backgrounds: current detection efforts and future prospects. Galaxies **10**(1), 34 (2022). https://doi.org/10.3390/galaxies10010034
36. K. Riles, Searches for continuous-wave gravitational radiation. Living Rev. Rel. **26**(1), 3 (2023). https://doi.org/10.1007/s41114-023-00044-3
37. Saleem M et al., The science case for LIGO-India. Class. Quantum Grav. **39**, 025004 (2022). https://doi.org/10.1088/1361-6382/ac3b99
38. N. van Remortel, K. Janssens, K. Turbang, Stochastic gravitational wave background: Methods and implications. Prog. Part. Nucl. Phys. **128**, 104003 (2023). https://doi.org/10.1016/j.ppnp.2022.104003
39. J. Veitch et al., Parameter estimation for compact binaries with ground-based gravitational-wave observations using the LALInference software library. Phys. Rev. D **91**(4), 042003 (2015). https://doi.org/10.1103/PhysRevD.91.042003
40. D. Wadekar et al., New black hole mergers in the LIGO-Virgo O3 data from a gravitational wave search including higher-order harmonics (2023 Dec), https://arxiv.org/abs/2312.06631

Open Access This chapter is licensed under the terms of the Creative Commons Attribution 4.0 International License (http://creativecommons.org/licenses/by/4.0/), which permits use, sharing, adaptation, distribution and reproduction in any medium or format, as long as you give appropriate credit to the original author(s) and the source, provide a link to the Creative Commons license and indicate if changes were made.

The images or other third party material in this chapter are included in the chapter's Creative Commons license, unless indicated otherwise in a credit line to the material. If material is not included in the chapter's Creative Commons license and your intended use is not permitted by statutory regulation or exceeds the permitted use, you will need to obtain permission directly from the copyright holder.

Multi-Detector Analyses for CCSN Neutrino Detection

Meriem Bendahman, Isabel Goos, Joao Coelho, Matteo Bugli, Alexis Coleiro, Sonia El Hedri, Thierry Foglizzo, Davide Franco, Jérôme Guilet, Antoine Kouchner, Raphaël Raynaud, and Yahya Tayalati

1 Introduction

Core-collapse supernovae (CCSNe) offer important insights into the dynamics of galaxies. Detecting the neutrino burst from a potential CCSN in the Milky Way before the visible explosion occurs is theoretically possible. Given the infrequency of galactic supernovae, it is crucial to promptly extract reliable CCSN localization information from neutrino data and transmit it efficiently to telescopes. Understanding the CCSN distance holds particular importance, as it could help ascertain whether the supernova took place in the dust-obscured regions behind the galactic center, influencing observation strategies [1]. However, untangling the impact of

---

M. Bendahman (✉)
Université Paris Cité, CNRS, Astroparticule et Cosmologie, Paris, France

Faculty of Sciences, Mohammed V University in Rabat, Rabat, Morocco
e-mail: mbendahman@km3net.de

I. Goos · J. Coelho · A. Coleiro · S. El Hedri · D. Franco · A. Kouchner
Université Paris Cité, CNRS, Astroparticule et Cosmologie, Paris, France

M. Bugli
Dipartimento di Fisica, Università di Torino, Torino, Italy

Université Paris-Saclay, Université Paris Cité, CEA, CNRS, AIM, Gif-sur-Yvette, France

T. Foglizzo · J. Guilet
Université Paris-Saclay, Université Paris Cité, CEA, CNRS, AIM, Gif-sur-Yvette, France

R. Raynaud
Université Paris Cité, Université Paris-Saclay, CEA, CNRS, AIM, Gif-sur-Yvette, France

Y. Tayalati
Faculty of Sciences, Mohammed V University in Rabat, Rabat, Morocco

Institute of Applied Physics, Mohammed VI Polytechnic University, Ben Guerir, Morocco

© The Author(s) 2026
Y. Tayalati, M. Gouighri (eds.), *The First African Conference on High Energy Physics*, Springer Proceedings in Physics 425,
https://doi.org/10.1007/978-3-031-88933-2_4

this distance from the effects of progenitor and neutrino properties on observations presents substantial challenges. Existing strategies for CCSN distance measurement aim to mitigate progenitor dependence but rely on the assumption of standard neutrino flavor conversion mechanisms [2, 3]. Approaches to constrain neutrino properties often hinge on energy measurements [4], a variable that may not be universally available across all experiments. Over the next two decades, the field of neutrino detectors sensitive to CCSNe will undergo a significant transformation, introducing new detectors sensitive to various combinations of neutrino flavors alongside traditional water Cherenkov detectors. This flavor complementarity could potentially break degeneracies between neutrino and CCSN properties even when the event energies are not known, thus improving the robustness of the CCSN distance measurements made by the Supernova Early Warning System (SNEWS) [5]. In this contribution, we introduce an algorithm designed to simultaneously constrain CCSN position, progenitor characteristics, and neutrino properties, utilizing only neutrino counting rates measured at large-scale next-generation experiments. We assess the algorithm's capability to locate CCSNe and characterize neutrinos with minimal information, while also evaluating its robustness in the presence neutrino two-body decays.

2 Methodology

In examining the effects of CCSN and neutrino properties on observations, we assess the anticipated neutrino rates across various current and upcoming experiments, employing a diverse range of supernova models and different flavor conversion mechanisms. This section outlines the methodology involved in this assessment and details the observables selected for characterizing CCSN.

CCSN Models In this analysis, we explore a collection of 149 progenitor models formulated and introduced in [6]. These models, established through one-dimensional simulations, encompass a broad spectrum of progenitor parameters such as masses, compactness, and metallicity, which are anticipated in CCSNe with an iron core. The purpose behind their creation was to identify observables with a weak dependence on the CCSN model, making them particularly well-suited for our study. The probability distribution for these models is defined using the Salpeter Initial Mass Function: $w(M) \propto M^{-2.35}$, M is the progenitor mass. The models are sampled using $w(M)$ as a weight.

Neutrino Experiments Future large-scale detectors will exhibit sensitivity to three neutrino flavor combinations. Water Cherenkov (WC) detectors will be responsive to electron antineutrinos, kiloton-scale liquid argon detectors to electron neutrinos, and detectors for Coherent Neutrino-Nucleus scattering (CEνNS) will be sensitive to the sum of all neutrino flavors. In the case of WC detectors, our considerations include Hyper-Kamiokande (HK), IceCube, and KM3NeT. For large liquid argon experiments we focus on DUNE's far detector, currently the most extensive project in this category. Additionally, for CEνNS detection experiments, we examine DarkSide-20k and a comparable but seven times larger project named ARGO.

Neutrino Flavor Conversion Models This analysis specifically delves into the initial 150 ms of a CCSN, where adiabatic MSW flavor transitions predominantly govern flavor-conversion mechanisms within the star. The MSW effect strongly depends on the mass ordering. Furthermore, we explore scenarios involving two-body decays of neutrinos. This BSM scenario represents one example of new physics phenomena which can mimic SM scenarios and introducing biases in distance and mass ordering estimates [7]. As an example, we consider the Dirac ϕ_0 scenario from [7], where the heaviest neutrino species decays into the lightest, introducing two parameters: the ratio $\bar{r}$ of the CCSN distance over the decay length and the branching ratio ζ to active neutrinos.

Analysis Pipeline Neutrino rates at various detectors are evaluated using the SNEWPY software [8]. SNEWPY has been modified to incorporate the described neutrino decay model, along with detection efficiency curves for DarkSide-20k and ARGO [9]. For IceCube and KM3NeT, Poissonian backgrounds are added, with rates of 1.5 and 3 MHz, respectively. More details can be found in [10].

Building Block Observables This analysis relies on observables with low or easily parameterizable dependencies on the CCSN model. Many of these observables, known as "standard candles" for CCSN distance measurements at single detectors, have been proposed in previous literature. While earlier time windows display weaker CCSN model dependence, the reduced statistical uncertainties for larger windows might lead to more precise measurements [2, 3]. A preliminary study is made to illustrate how to separate mass ordering using early time windows by examining ratios between the rates described above and those measured during the early accretion phase, specifically 100–150 ms after the beginning of the CCSN. This choice aims to mitigate the residual model-dependence of early neutrino rates for CCSN distance measurements. The associated observables, referred to as f_Δ and illustrated in Fig. 1, exhibit quasi-linear dependence on early neutrino rates.

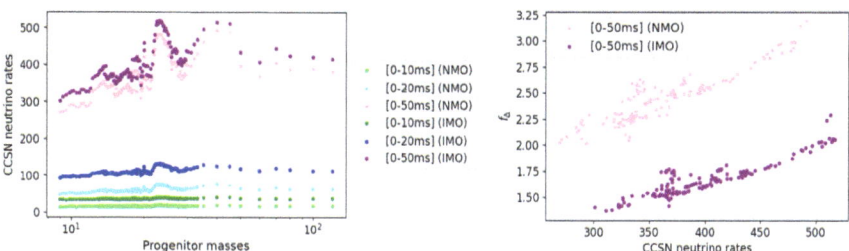

Fig. 1 Left: CCSN neutrino rates in Hyper-Kamiokande for a CCSN at 10 kpc as a function of progenitor masses in $M_\odot$, for the first 10 ms (green), 20 ms (blue), and 50 ms (purple) postbounce. Right: f_Δ as a function of the CCSN neutrino rates in the first 50 ms. NMO and the IMO, are presented with light and dark colors respectively

$$f_\Delta(\Delta t) = \frac{N(100-150\,\text{ms})}{N(\Delta t)} \approx \alpha N(\Delta t) + \beta \tag{1}$$

where Δt is the time window and N the neutrino rate.

In the study by [2], a value of Δt was set at 50 ms.

3 Likelihood Analysis

Considering a set of neutrino detectors, we establish the following likelihood function:

$$\log \mathcal{L}(\{\mathcal{O}_{\text{obs}}\}|d, M, \bar{r}, \zeta, \text{MO}) = \sum_i \log P[N_i(10\,\text{ms})] + \log P[N_i(10\text{--}20\,\text{ms})]$$
$$+ \log P[N_i(20\text{--}50\,\text{ms})] + \log P[N_i(100\text{--}150\,\text{ms})] \tag{2}$$

The index i represents the individual neutrino detectors, M and d represent the progenitor mass, and the supernova distance, respectively. The parameters $(\bar{r}, \zeta)$ correspond to the neutrino decay parameters. P denotes the Poisson probability distribution for observing a specific count of events:

$$P(N_i) = \frac{\lambda_i^{N_i} e^{-\lambda_i}}{N_i!}, \tag{3}$$

$\lambda_i(d, M, \bar{r}, \zeta, \text{MO})$ is the expected value of N_i considering CCSN and neutrino properties, $\{\mathcal{O}_{\text{obs}}\}$ denotes the collection of measurements conducted for various detectors. When constraining a specific parameter, such as the CCSN distance, the remaining parameters are regarded as nuisance parameters Ξ. In this context, we establish a profile likelihood:

$$\mathcal{L}_{\text{prof}}(\{\mathcal{O}_i\}|\theta) = \max_\Xi \mathcal{L}(\{\mathcal{O}_i\}|\theta, \Xi) \tag{4}$$

which can be used for either parameter fitting or hypothesis testing. To optimize the likelihood over $(\bar{r}, \zeta)$, a regular grid is considered, where: $\Delta\bar{r} = 0.05$ and $\Delta\zeta = 0.1$. More details can be found in [10].

4 Mass Ordering Determination

We employ the likelihood specified in Eq. (2) to deduce the neutrino Mass Ordering (MO) without knowledge of the CCSN distance. Initially assuming that neutrino properties follow the Standard Model (SM), we maximize the likelihood over

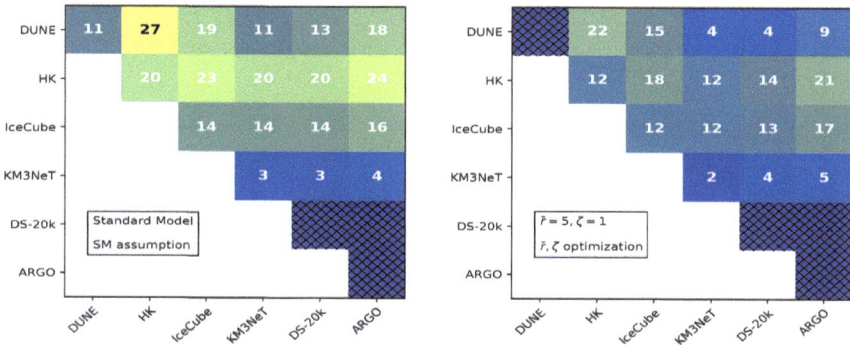

Fig. 2 3σ distance horizon (in kpc) indicating the rejection for the IMO when the NMO is the true mass ordering. The individual detectors are shown in the diagonal, off-diagonal elements display pairs of detectors. The cross-hatched regions identify experiments or pairs of experiments lacking sensitivity to the mass ordering, such as DUNE in the presence of neutrino decays, and CEνNS experiments like ARGO and DarkSide-20k. Left: SM scenario utilizing a SM likelihood. Right: Beyond Standard Model scenario with $\bar{r} = 5$, $\zeta = 1$, with $\bar{r}$ and ζ treated as degrees of freedom in the likelihood

distances and supernova models for each mass ordering (MO) hypothesis given a particular measurement. Subsequently, we compute the p-value for a IMO or NMO measurement, using the ratio of IMO and NMO likelihoods as test statistic defined as:

$$t(\{\mathcal{O}_{\text{obs}}\}) = \frac{\max_{d,M,\bar{r},\zeta} [\mathcal{L}(\{\mathcal{O}_{\text{obs}}\}|d, M, \bar{r}, \zeta, \text{NMO})]}{\max_{d,M,\bar{r},\zeta} [\mathcal{L}(\{\mathcal{O}_{\text{obs}}\}|d, M, \bar{r}, \zeta, \text{IMO})]} \quad (5)$$

The likelihoods are optimised over both CCSN distance and the CCSN models. In the scenario of the Standard Model, the parameters are set to $\bar{r} = 0$ and $\zeta = 1$. However, if the potential for neutrino decays is taken into consideration, the likelihoods are optimised over the $\bar{r}$ and ζ parameters. To assess the probability distribution of t under a specific mass ordering hypothesis, we generate potential observations by sampling from a prior probability distribution described in [10]. Figure 2 illustrates the 3 σ distance horizon for rejecting the IMO based on the distribution of the test statistic t under the NMO case, considering both single and pairs of experiments. Combining Hyper-Kamiokande and IceCube extends the distance horizon from 20 kpc, when considering Hyper-Kamiokande alone, to 23 kpc. Notably, the most significant enhancement occurs when combining Hyper-Kamiokande with DUNE, resulting in a 27 kpc horizon that encompasses the entire galaxy. Likewise, for the IMO, the 3 distance horizon extends from 11 kpc to 18 kpc when combining DUNE with ARGO and from 14 kpc to 19 kpc when pairing DUNE with IceCube. We then assess the ability of neutrino experiments to differentiate between the NMO and the IMO scenarios, considering the presence of neutrino decays. The right panel of Fig. 2 displays the corresponding distance horizons for rejecting the Standard Model in the IMO. In this analysis, we simulated

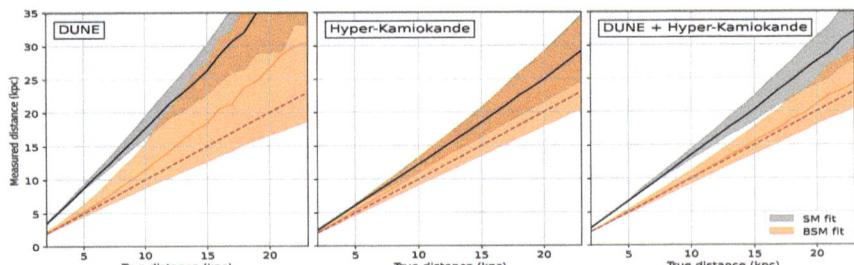

Fig. 3 Median values and 90% confidence intervals on the measured CCSN distance are presented as functions of the true distance, for a $11 M_\odot$ progenitor, assuming the IMO and a neutrino decay model of $\bar{r} = 5, \zeta = 1$. DUNE, HK and their combination are presented. The grey band is for the SM hypothesis, while the orange band is a fit where $\bar{r}$ and ζ are optimized alongside the other parameters

an observation for which $\bar{r} = 5, \zeta = 1$, but we vary $\bar{r}$ and ζ when we optimize the likelihood. For IMO rejection, the largest distance horizons are achieved by combining DUNE with HK (22 kpc) and by combining HK with ARGO (21 kpc).

5 Measuring Supernova Distances

To investigate the effect of neutrino decays on CCSN distance measurements, we examine a representative measurement for an 11 $M_\odot$ progenitor, considering a specific neutrino decay scenario ($\bar{r} = 5$ and $\zeta = 1$). In Fig. 3, the measured CCSN distance and the corresponding 90% confidence interval are depicted as functions of the true distance for the IMO scenario with ($\bar{r} = 5, \zeta = 1$). In Fig. 3, the measured CCSN distance and the corresponding 90% confidence interval are depicted as functions of the true distance for the IMO scenario with ($\bar{r} = 5, \zeta = 1$) mentioned earlier. This analysis is conducted for the DUNE and HK detectors individually, as well as for their combined results. In this figure, the distance estimated using the SM-only likelihood is compared to the distance obtained by allowing $\bar{r}$ and ζ to vary. For both DUNE and HK, a noticeable bias in the distance measurement is evident under the SM assumption. This bias can be, to some extent, rectified by incorporating $\bar{r}$ and ζ parameters in the CCSN distance fitting process. In that case, when combining DUNE and HK, the 90% C.L. uncertainties become comparable to the ones obtained under the SM assumption.

6 Conclusion

In this contribution, we demonstrated that with a method using a minimal set of observables, and exploiting the capabilities of the next generation of neutrino experiments, the CCSN alert systems impose constraints on CCSN properties even when the energies of the detected events are unknown. Taking neutrino decays an example, we showed how new physics in the neutrino sector can introduce notable biases into CCSN distance estimates. We showed that the neutrino mass ordering can be determined independently using the method described in Sect. 4. We also have demonstrated that when new physics impacts neutrino fluxes in a flavor-dependent manner, it becomes possible to correct distance estimates while maintaining control over uncertainties. This correction can be achieved through the combination of flavor-complementary neutrino detectors.

Acknowledgments This work is supported by LabEx UnivEarthS (ANR-10-LABX-0023 and ANR-18-IDEX-0001) and Paris Region (DIM ORIGINES). We thank Gwenhaël de Wasseige, Maria Cristina Volpe and Joachim Kopp for interesting and helpful discussions. This project has received funding from the European Union's Horizon Europe research and innovation programme under the Marie Skłodowska-Curie grant agreement No 101064953 (GR-PLUTO). Joao Coelho, Alexis Coleiro, Sonia El Hedri, Davide Franco, Isabel Goos and Antoine Kouchner are supported by Centre National de la Recherche Scientifique (CNRS). The Moroccan Ministry of Higher Education, Scientific Research and Innovation is acknowledged.

References

1. K. Nakamura et al., Multimessenger signals of long-term core-collapse supernova simulations: synergetic observation strategies. Mon. Natl. R. Astron. Soc. **461**, 3296–3313 (2016). https://doi.org/10.1093/mnras/stw1453
2. M. Segerlund et al., Measuring the distance and mass of galactic core-collapse supernovae using neutrinos. astro-ph.HE (2021). https://doi.org/10.48550/arXiv.2101.10624
3. M. Kachelrieß et al., Measuring the distance and mass of galactic core-collapse supernovae using neutrinos. Phys. Rev. D **71**(16), 063003 (2005). https://doi.org/10.1103/PhysRevD.71.063003
4. X.-R. Huang et al., Bayesian inference of supernova neutrino spectra with multiple detectors. J. Cosmol. Astropart. Phys., 040 (2023). https://doi.org/10.1088/1475-7516/2023/09/040
5. S. Al Kharusi et al., SNEWS 2.0: a next-generation supernova early warning system for multimessenger astronomy. New J. Phys. **23**, 031201 (2021). https://doi.org/10.1088/1367-2630/abde33
6. T. Sukhbold et al., Core-collapse supernovae from 9 to 120 solar masses based on neutrino-powered explosions. Astrophys. J. **821**, 38 (2016). https://doi.org/10.3847/0004-637X/821/1/38
7. A. de Gouvea et al., Impact of neutrino decays on the supernova neutronization-burst flux. Phys. Rev. D **101**, 043013 (2020). https://doi.org/10.1103/PhysRevD.101.043013
8. A. Baxter et al., SNEWPY: A data pipeline from supernova simulations to neutrino signals. J. Open Source Softw. **6**(67), 3772 (2021). https://doi.org/10.21105/joss.03772
9. The DarkSide-20k collaboration et al., Sensitivity of future liquid argon dark matter search experiments to core-collapse supernova neutrinos. J. Cosmol. Astropart. Phys. **03**, 043 (2021). https://doi.org/10.1088/1475-7516/2021/03/043
10. M. Bendahman et al., Prospects for realtime characterization of core-collapse supernova and neutrino properties. J. Cosmol. Astropart. Phys. **02**, 008 (2024). https://doi.org/10.1088/1475-7516/2024/02/008

Open Access This chapter is licensed under the terms of the Creative Commons Attribution 4.0 International License (http://creativecommons.org/licenses/by/4.0/), which permits use, sharing, adaptation, distribution and reproduction in any medium or format, as long as you give appropriate credit to the original author(s) and the source, provide a link to the Creative Commons license and indicate if changes were made.

The images or other third party material in this chapter are included in the chapter's Creative Commons license, unless indicated otherwise in a credit line to the material. If material is not included in the chapter's Creative Commons license and your intended use is not permitted by statutory regulation or exceeds the permitted use, you will need to obtain permission directly from the copyright holder.

Towards a Finite AdS$_3$ Topological Gravity Landscape

Rajae Sammani , **Youssra Boujakhrout, El Hassan Saidi, Rachid Ahl Laamara, and Lalla Btissam Drissi**

1 Introduction

Since the usual tools and techniques of quantum field theory failed to construct a renormalisable model of quantum gravity, it was suggested that the consistency, or the lack of, is intrinsically due to gravity. There must be an additional set of criteria, proper to gravity, that are required to be verified in order for any quantum gravitational model to be consistent independently of a particular UV completion. Though, what are these criteria? How many of them are there? And most importantly, how can we determine them?

This is the basic aim of the swampland program (SP) [1], establishing the various consistency conditions for effective field theories coupled to gravity by identifying, formulating, testing, and refining the swampland conjectures. So far, we have several swampland conjectures, each constraining a different parameter of the effective quantum gravitational models. The one of interest is the Landscape finiteness conjecture [2]. It was first established by Vafa [2] and expressed by un upper bound on the number of massless modes and was tested in various string theory models with different dimensions and supercharges.

In this work, we further test the validity of the finiteness conjecture in 3d. However, in this standard case, gravity is topological and might be too trivial for the swampland analysis. Instead, we consider the addition of a negative cosmological constant to work in a curve AdS$_3$ space with a CFT$_2$ boundary providing therefore

---

R. Sammani (✉) · Y. Boujakhrout · E. H. Saidi · R. Ahl Laamara · L. B. Drissi
LPHE-MS, Science Faculty, Mohammed V University in Rabat, Rabat, Morocco

Centre of Physics and Mathematics, CPM, Rabat, Morocco
e-mail: rajae_sammani@um5.ac.ma; youssra_boujakhrout@um5.ac.ma; e.saidi@um5r.ac.ma; r.ahllaamara@um5r.ac.ma; ldrissi@fsr.ac.ma

conformal tools at our disposal. Also, the theory has a gauge formulation given by the Chern-Simons (CS) formulation based on the gauge group $sl(2)_L \times sl(2)_R$, which draws the line for several types of higher spin generalisation by principally embedding the fundamental $sl(2)$ in various higher spin gauge symmetries. Subsequently, in order to derive the swampland finiteness constraint in the new setting, we must define the AdS$_3$ Landscape, study the consistency of the model by insuring the cancellation of any emerging anomalies then compute the upper bound on the rank.

2 Extended AdS3 Asymptotic Symmetry

The three dimensional Einstein-Hilbert action with a negative cosmological constant can be formulated as a gauge theory given by the difference between two CS actions based on the $sl(2)_L \times sl(2)_R$ symmetry as follows [3]

$$-\frac{1}{16\pi G_N} \int d^3x \sqrt{g} \left(R - 2\Lambda\right) = S_{CS}\left(A_L\right) - S_{CS}\left(A_R\right) \tag{1}$$

where

$$S_{CS}(A) = \frac{k}{4\pi} \int tr \left(AdA + \frac{2}{3}A^3\right) \tag{2}$$

Unfortunately, without imposing boundary conditions, the action is anomalous as it has a defective variational principle. In fact,

$$\delta S_{CS}[A] = \frac{k}{2\pi} \int_{\mathcal{M}_{3D}} tr\left[\delta A \left(dA + A^2\right)\right] + \frac{k}{4\pi} \int_{\partial \mathcal{M}_{3D}} tr(\delta A A) \tag{3}$$

using the on-shell equation $F = dA + A^2 = 0$, the bulk vanishes and we are left with the boundary term. Switching into a radial gauge, the 1-form potential $A = A_t dt + A_\varphi d\varphi$ are expressed like $a_t dt + a_\varphi d\varphi$, the variation therefore becomes

$$\delta S_{CS} = \frac{k}{4\pi} \int_{\partial \mathcal{M}_{3D}} dt d\varphi \sqrt{|h|} \left(\delta a_t a_\varphi - \delta a_\varphi a_t\right) \tag{4}$$

with h the determinant of the boundary metric $h_{\alpha\beta}(t, \varphi)$. In order to restore a healthy variational principle, it is too restraining to require both δa_t and δa_φ to cancel. Instead, one can add a boundary term S^{bnd} given by

$$S^{bnd} = \frac{k}{4\pi} \int_{\partial \mathcal{M}_{3D}} dt d\varphi \sqrt{|h|} tr(a_t a_\varphi) \tag{5}$$

the new variation δS_{tot}^{CS} becomes

$$\delta S_{tot}^{CS} = \delta(S_{CS} + S_{CS}^{bnd}) = \frac{k}{2\pi} \int_{\partial M_{3D}} dt d\varphi \sqrt{|h|} tr\left[\delta a_t a_\varphi\right] \tag{6}$$

With only $\delta a_t = 0$, the gauge anomaly will cancel and the non-variation is restored.

However, the theory's physical states are set in a representation of the loop algebra [4], and the physics is very sensitive to the choice of the boundary conditions. A choice as $\delta a_t = 0$ affects the resulting asymptotic symmetry algebra (ASA). The derivation of the ASA follows from the computation of the boundary charge as

$$\begin{aligned} \delta_\zeta Q &= \tfrac{k}{2\pi} \int tr\left[\zeta\left(\delta a_\varphi d\varphi + \delta a_t dt\right)\right] \\ &= \tfrac{k}{2\pi} \int tr\left(\zeta \delta a_\varphi\right) d\varphi \equiv \delta_\zeta Q^\varphi \end{aligned} \tag{7}$$

where the choice of boundary conditions $\delta a_t = 0$ omits half of the asymptotic symmetry. Instead, if one takes both components as non-vanishing $\delta a_t \neq 0$ and $\delta a_\varphi \neq 0$, the resulting asymptotic affine algebra is an extension of the usual one [4]. In fact, by considering

$$\begin{aligned} \delta_\zeta Q &= \tfrac{k}{2\pi} \int tr\left[\zeta\left(\delta a_\varphi d\varphi + \delta a_t dt\right)\right] \\ &\equiv \delta_\zeta Q^\varphi + \delta_\zeta Q^t \end{aligned} \tag{8}$$

which gives

$$Q_\zeta^\alpha = \int tr\left(\zeta^0 a_\alpha^0 + \zeta^+ a_\alpha^- + \zeta^- a_\alpha^+\right) d\alpha \tag{9}$$

with boundary connections expanded in the $sl(2)$ basis as:

$$a_\varphi = \frac{-2\pi}{k}\left[a_\varphi^+ K_+ - 2a_\varphi^0 K_0 + a_\varphi^- K_-\right] \tag{10}$$

$$a_t = \frac{-2\pi}{k}\left[a_t^+ K_+ - 2a_t^0 K_0 + a_t^- K_-\right] \tag{11}$$

The ASA algebraic brackets of the left sector are therefore given by $\delta_\zeta a_\varphi = \{a_\varphi, Q_\zeta^\varphi\}_{DB}$ and $\delta_\zeta a_t = \{a_t, Q_\zeta^t\}_{DB}$ where $\delta_\zeta a = d\zeta + [a, \zeta]$. The extended ASA for the left sector is given by two copies of the affine $sl(2)_{k_L}$ algebra given by

$$\left\{a_\varphi^a(t,\varphi), a_\varphi^b(t,\bar\varphi)\right\}_{DB} = \delta(\varphi - \bar\varphi)(a-b) a_\varphi^{a+b}(t,\varphi) + \frac{k_L}{2\pi}\kappa^{ab}\frac{\partial}{\partial\varphi}\delta(\varphi - \bar\varphi) \tag{12}$$

$$\left\{a_t^a(t,\varphi), a_t^b(\bar t,\varphi)\right\}_{DB} = \delta(t-\bar t)(a-b) a_t^{a+b}(t,\varphi) + \frac{k_L}{2\pi}\kappa^{ab}\frac{\partial}{\partial t}\delta(t-\bar t) \tag{13}$$

A similar analysis for the right handed sector results in additional two copies of the affine $sl(2)_{k_R}$. The extended algebra (12) can also be derived using the variational principle by first computing the anomalous currents then the associated algebra. This method was reported in [5].

3 Anomaly Cancellation and Emerging Strings

Even though the asymptotic symmetry (12, 13) is more encompassing, the choice of boundary conditions is defective as it leads to a gauge anomaly exhibited by the violation of the current conservation given by

$$\delta \mathcal{S}_{tot}^{CS} = \int_{\partial \mathcal{M}_{3D}} dt d\varphi \sqrt{|h|} J_{\varphi c} \left(\delta a_t^c\right) \Rightarrow J_{\varphi c} = \delta \mathcal{S}_{tot}^{CS}/\left(\delta a_t^c\right) = \frac{k}{2\pi} \kappa_{cb} a_\varphi^b \qquad (14)$$

where $\nabla J_{\varphi c} \neq 0$. Our next goal is to cancel this anomaly without the restricting the boundary charge. In other words, we need to neutralize $\nabla J_{\varphi c}$ without imposing $\delta a_\varphi = 0$. Recall that the bulk CS theory (2) is dual to a chiral WZW model at the boundary

$$\mathcal{S}_{wzw}^L (g_L) = \frac{k}{4\pi} \int_{\partial M} dt d\varphi Tr \left[\left(g_L^{-1} \partial_t g_L\right)\left(g_L^{-1} \partial_\varphi g_L\right)\right] + \frac{k}{12\pi} \int_M Tr \left(g_L^{-1} dg_L\right)^3 \qquad (15)$$

And since the edge term can be interpreted as the Polyakov action of a bosonic string in the conformal gauge $\int_{\partial M} d^2 z \sqrt{|-h|} h^{\alpha\beta} G_{ab} \partial_\alpha X^a \partial_\beta X^b$, where $g(x) = e^{K_a X^a(x)}$, we can therefore introduce a boundary bosonic string via a 2D action coupled to an external gauge field a_t as

$$\mathcal{S}_{string} = \frac{k'}{4\pi} \int_{\partial M} dt d\varphi Tr \left[\left(g_L^{-1} \partial_t g_L\right)\left(g_L^{-1} \partial_\varphi g_L\right) + 2 g_L^{-1} \partial_\varphi g_L a_t\right]$$
$$+ \frac{k'}{12\pi} \int_M Tr \left(g_L^{-1} dg_L\right)^3 \qquad (16)$$

with a anomalous current given by $J_\varphi = \delta \mathcal{S}_{tot}^{CS}/(\delta a_t) = \frac{k'}{2\pi} g_L^{-1} \partial_\varphi g_L = \frac{k'}{2\pi} a_\varphi$. Considering the total action $\mathcal{S}_{tot}^{CS} + \mathcal{S}_{string}$ with $k' = -k$, we get two counteracting currents and the resulting model has therefore an anomaly free boundary.

After generalising the argument for the right sector, the AdS$_3$ gravity with extended boundary conditions has thus an ASA given by four copies of the sl(2)$_k$ affine algebra instead of the usual two with a non-anomalous 2D boundary. The full computation will be reported in a future work.

4 Towards a Finite AdS$_3$ Landscape

Another remarkable property of the AdS$_3$ gravity model is the ability to couple gravity to a finite number of higher spin fields without forgoing its consistency by simply promoting the sl(2) to higher dimensional symmetries via the principle embedding of [6]. One can hence construct an AdS$_3$ Landscape as a set of the higher spin AdS$_3$ gravity gauge symmetries [7].

	AdS$_3$ Landscape			
A_N	$SL(N,R)_L \times SL(N,R)_R$ $SU(1,N-1)_L \times SU(1,N-1)_R$ $SU\left(\frac{N+1}{2},\frac{N+1}{2}\right)_L \times SU\left(\frac{N+1}{2},\frac{N+1}{2}\right)_R$	C_N		$Sp(2N,R)_L \times Sp(2N,R)_R$ $Sp(P,2N-P)_L \times Sp(P,2N-P)_R$ $Sp(N,N)_L \times Sp(N,N)_R$
B_N	$SO(N,1+N)_L \times SO(N,1+N)_R$ $SO(1,2N)_L \times SO(1,2N)_R$ $SO(P,2N+1-P)_L \times SO(P,2N+1-P)_R$	G_2 F_4		$G_{2(2)} \times G_{2(2)}$ $F_{4(4)} \times F_{4(4)}$ $F_{4(-20)} \times F_{4(-20)}$
D_N	$SO(N,N)_L \times SO(N,N)_R$ $SO(N-1,N+1)_L \times SO(N-1,N+1)_R$ $SO(1,2N-1)_L \times SO(1,2N-1)_R$	E_8		$E_{8(8)} \times E_{8(8)}$ $E_{8(-24)} \times E_{8(-24)}$
E_6	$E_{6(6)} \times E_{6(6)}$ $E_{6(2)} \times E_{6(2)}$ $E_{6(-14)} \times E_{6(-14)}$ $E_{6(-26)} \times E_{6(-26)}$	E_7		$E_{7(7)} \times E_{7(7)}$ $E_{7(-5)} \times E_{7(-5)}$ $E_{7(-25)} \times E_{7(-25)}$

(17)

The finiteness of the Landscape is bounded to the finiteness of the higher spin families. One has to derive a consistency constraint on the infinite families and compute un upper bound on their rank r in order for the AdS3 Landscape to be finite in accordance with the Swampland conjecture. For instance, in the left sector, the currents algebras are bounded by the left central charge as follows: $\sum_i c_{\mathcal{G}_{iL}}(r) < c_L$. The mapping of the central charges to each one of the obtained copies (12–13) in addition to the computation of the explicit bounds for split reals forms of (17), are investigated in [5].

5 Conclusion

The successfulness of the Landscape analysis regarding the finiteness conjecture on topological 3D gravitational theories is enticing, as it prompts consideration of the possible implications of other swampland criteria. In future work, we investigate the AdS distance conjecture and the weak gravity constraint [8, 9]. We also study the associated higher spin black holes solutions for the various real forms of the Landscape symmetries [10].

References

1. T.D. Brennan, F. Carta, C. Vafa, The string landscape, the swampland, and the missing corner. Preprint. arXiv:1711.00864 (2017)
2. H.C. Kim, G. Shiu, C. Vafa, Branes and the Swampland. Phys. Rev. D **100**(6), 066006 (2019)
3. A. Achucarro, P.K. Townsend, A Chern-Simons action for three-dimensional anti-de Sitter supergravity theories. Phys. Lett. B **180**(1–2), 89–92 (1986)
4. D. Grumiller, M. Riegler, Most general AdS3 boundary conditions. J. High Energy Phys. **2016**, 23 (2016)
5. R. Sammani, Y. Boujakhrout, R.A. Laamara, L.B. Drissi, Finiteness of 3D higher spin gravity landscape. Class. Quantum Grav. **41**(21), 215012 (2024)
6. A. Campoleoni, S. Fredenhagen, S. Pfenninger, S. Theisen, Asymptotic symmetries of three-dimensional gravity coupled to higher-spin fields. J. High Energy Phys. **11**, 1–36 (2010)
7. R. Sammani, Y. Boujakhrout, E.H. Saidi, R.A. Laamara, L.B. Drissi, Higher spin AdS3 gravity and Tits-Satake diagrams. Phys. Rev. D **108**(10), 106019 (2023)
8. R. Sammani, E.H. Saidi, Higher spin swampland conjecture for massive AdS$_3$ gravity. Preprint. arXiv:2406.09151 (2024)
9. M. Charkaoui, R. Sammani, E.H. Saidi, R.A. Laamara, Asymptotic weak gravity conjecture in M-theory on K3 × K3. Prog. Theor. Exp. Phys. **7**, 073B08 (2024)
10. R. Sammani, E.H. Saidi, Black flowers and real forms of higher spin symmetries. J. High Energy Phys. **2024**, 44 (2024)

Open Access This chapter is licensed under the terms of the Creative Commons Attribution 4.0 International License (http://creativecommons.org/licenses/by/4.0/), which permits use, sharing, adaptation, distribution and reproduction in any medium or format, as long as you give appropriate credit to the original author(s) and the source, provide a link to the Creative Commons license and indicate if changes were made.

The images or other third party material in this chapter are included in the chapter's Creative Commons license, unless indicated otherwise in a credit line to the material. If material is not included in the chapter's Creative Commons license and your intended use is not permitted by statutory regulation or exceeds the permitted use, you will need to obtain permission directly from the copyright holder.

Maximinext Surfaces: Enhancements in Island Entropy

Safae Tariq, R. Ahl Laamara, and E. H. Saidi

1 Introduction

Black holes are quantum gravitational systems identified with holographic entanglement entropy in the framework of AdS/CFT [1]. Bekenstein [2] defined entropy as proportional to the horizon area of a black hole. Hawking [3] developed this concept in semi-classical gravity, which is known as Bekenstein-Hawking entropy. From a holographic standpoint, the flame is caught by Ryu-Takayanagi (RT) proposal [4] which suggest that the entropy is proportional to a minimal surface area and HRT proposal [5]. As an aside, the black hole entropy can be written in terms of quantum extremal surfaces (QES) [6] and quantum maximin surfaces (QMM) [7]. We suggest a new surface to compute the von Neumann entropy in the bulk called maximinext (MME) surface.

In this work, we recast the properties of classical surfaces and quantum surfaces. We investigate the holographic entanglement entropy formulas in terms of MME surface. We establish the island entropy using MME surfaces following [8].

The outline of the paper is as follows: In Sect. 2, we revisit some aspects of classical surfaces and quantum surfaces such as RT, HRT, QES and QMM. Section 3 is fully reserved for implementing new island entropy proposal utilizing maximinext surfaces. In Sect. 4, we give a summary of our work and discuss our main results.

S. Tariq (✉) · R. Ahl Laamara · E. H. Saidi
LPHE-MS, Science Faculty, Mohammed V University in Rabat, Rabat, Morocco

Centre of Physics and Mathematics, CPM, Rabat, Morocco
e-mail: safae.tariq@um5r.ac.ma

2 Gravitational Fine-Grained Entropy

In this section, we revisit the aspects of holographic entanglement entropy written in terms of quantum extremal surfaces (QES) [6] and quantum maximin surfaces (QMM) [7]. We suggest a new kind of surfaces called maximinext surface.

2.1 Some Aspects of Classical Surfaces

In semi-classical gravity, the black hole entropy is described with the horizon area of the black hole known as the Bekenstein-Hawking entropy [2, 3]

$$S = \frac{Area}{4G_N \hbar} \quad (1)$$

which does not satisfy the second law of thermodynamics, for that von Neumann entropy adjusted to be the quantum correction to this entropy. Then, combining both blocks we get the generalized entropy $S_{gen} = \frac{Area}{4G_N \hbar} + S_{vN}$. However, in gravitational theories S_{vN} can be computed via classical surfaces as RT [4], HRT [5] and maximin surfaces [9], or via quantum surfaces such as quantum extremal surfaces [6] and quantum maximin surfaces [7].

The Ryu-Takayanagi (RT) [4] proposal focuses on investigating the entanglement entropy in CFT under AdS/CFT duality. Assuming that our system is divided into two subsystems, A and B, the whole Hilbert space can be acquired as a direct product of two spaces $\mathcal{H}_{tot} = \mathcal{H}_A \otimes \mathcal{H}_B$ for A and B. If we have an observer who can only access the subsystem, A, the reduced density matrix $\rho_A = Tr_B \rho_{tot}$ can be obtained by tracing over the Hilbert space $\mathcal{H}_B$. Then, the entropy takes the form

$$S_A = -tr_A(\rho_A \log \rho_A) \quad (2)$$

The above entanglement entropy satisfies the following properties [4]: (1) Equality of entropies of subsystems A and B, (2) subadditivity, and (3) strong subadditivity.

In AdS/CFT framework, the CFT entanglement entropy is defined by the minimal surface area of the black hole horizon as

$$S_A = \frac{Area\ of\ \gamma_A}{4G_N^{(d+2)}} \quad (3)$$

Here, γ_A is the d dimensional static minimal surface in AdS_{d+2} with boundary ∂A, and $G_N^{(d+2)}$ is the d+2 dimensional Newton constant.

Subsequently, RT proposal [4] is not dynamically developed in spacetime, for that HRT (Hubeny, Rangamani and Takayanagi) [5] connects the von Neumann

Fig. 1 Minimal area extremal surface σ_R homologous to a boundary region R

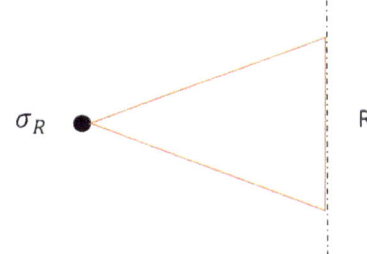

entropy of a boundary subregion R to the size of a minimal area extremal surface σ_R homologous to boundary region R as suggested in Fig. 1.

Physically, the increasing or decreasing of the area σ_R is specified by the sign of $\kappa^a n_a$ along some vector field n_a. To elucidate, instead of minimizing the area, HRT suggests finding the extremal area,

$$S_A = ext(\frac{A}{4}) \qquad (4)$$

To include quantum corrections, we generalize quantum mechanically these classical surfaces in the following subsection.

2.2 Quantum Generalizations of Entanglement Entropy

Here, we emphasize the quantum generalizations of the HRT [5] and maximin surfaces [9] as the corresponding bulk entropies of radiation entropy. The fine-grained entropy can also be given by a formula that involves a generalized entropy. We suggest a new von Neumann entropy by maximinextremizing the generalized entropy functional.

Engelhardt and Wall [6] proposed a new approach to finding the minimal surface in a gravitational system. The resulting surface is called the quantum extremal surface which is a quantum deformation of the extremal surface. However, extreme surfaces that minimize in spatial direction but maximize in time direction the generalized entropy [6] are

$$S = \min_{\chi}\{ext_{\chi}[\frac{Area}{4G_N} + S_{bulk}]\} \qquad (5)$$

As an aside, QES has featured various boundary conditions [6] on the asymptotic boundary, of which we have three types (1) absorbing, (2) reflecting, and (3) open. The above entropy formula can be written with another kind of surfaces mentioned by Wall [9] and developed by Akers et al. [7] called quantum maximin surfaces defined as for every Cauchy slice containing R we find the minimal S_{gen} surface homologous to R. We then look for the maximum surface among all of these minima

$$S_{vN}[\rho_R] = \max_{Cauchy} \min_{\chi}[S_{gen}(\chi)] \qquad (6)$$

This quantum maximin surfaces [7] must satisfy some criteria, such as stability and existence.

Furthermore, we investigate a new rule for computing the fine-grained entropy using Maximinext surfaces. Starting by identifying a specific region in spacetime called Wheeler-de Witt patch (WdW) $\mathcal{W}$ [8]. Then, we choose a Cauchy slice C for $\mathcal{W}$ where x defined as an extremal surface. We minimize x over η in the Cauchy slice C; $\eta \subset C$, and maximize the minext surface. Moreover, by maximinextremizing the generalized entropy functional we get the new formula of von Neumann entropy in the WdW patch that informs us about the whole spacetime.

$$S_{vN}[\rho_R] = \max_C \min_\eta ext_x [\frac{Area}{4G_N} + S_{bulk}] \qquad (7)$$

The above entropy is special because it combine both the maximin and quantum extremal surfaces.

3 Island Holographic Entanglement Entropy

In this section, we inspect the island entropy in JT gravity theory with matter for a black hole evaporating in a bath that is the CFT_2. By doing so, the radiation region is related to the region behind the horizon under additional dimension via ER=EPR theory [1]. We suggest a new way to compute the radiation entropy utilizing the maximinext surface outlined in the last section.

Starting with identifying the quantum extremal surfaces by extremizing the generalized entropy functional but here we include the island contribution. When we include such islands we will have to pay a price due to the boundary area of the island, which correspond to the area term in the gravity theory. The prescription is that the actual entropy of some region A in a quantum field theory is given by extremizing a generalized entropy-like functional over islands $\mathcal{I}_g$ followed by a minimization over all extrema [1]:

$$S(A) = \underset{\mathcal{I}_g}{Min} \underset{\mathcal{I}_g}{Ext} \left[S_{eff}(A \cup \mathcal{I}_g) + \frac{Area[\partial \mathcal{I}_g]}{4G_N} \right] \qquad (8)$$

where $Area[\partial \mathcal{I}_g]$ corresponds to the boundary area of the island. We call the islands $\mathcal{I}_g$ that extremize this functional quantum extremal islands. Here, we have a state $\rho_{A \cup \mathcal{I}_g}^{eff}$ which is a semiclassical gravity state.

The island gravitational fine-grained entropy might be written using the maximinext (MME) proposal as suggested in the last section as

$$S(A) = \underset{\mathcal{J}_g}{Max}\underset{\mathcal{N}}{Min}\underset{\chi}{Ext}\left[S_{eff}(A \cup \mathcal{J}_g) + \frac{Area[\partial \mathcal{J}_g]}{4G_N}\right] \quad (9)$$

Here, we extremize over the QES χ which is the boundary of the island and minimize over a subregion of the island $\mathcal{N}$; $\mathcal{N} \subset \mathcal{J}_g$. Then, we maximize the minext over the total island region $\mathcal{J}_g$ sited behind the horizon. This equation maximinextremize the functional $S_{eff}(A \cup \mathcal{J}_g) + \frac{Area[\partial \mathcal{J}_g]}{4G_N}$, which obeys the generalized second law of thermodynamics since is composed of two blocks; the first corresponds to the quantum correction, and the second represents thermodynamic entropy.

4 Conclusion

In this paper, we recasted the computation of entanglement entropy using RT/HRT proposals. Then, we outlined the derivation of holographic entanglement entropy utilizing quantum extremal surfaces and quantum maximin surfaces. We suggested new maximinext surfaces to determine new version of black hole entropy. Moreover, we established new release of island entropy proposal founded by MME surfaces.

In Sect. 2, we revisited the quantum generalization of HRT and maximin surfaces used in computing the holographic entanglement entropy. We investigated new surface called maximinext surface to reformulate the HEE entropy. In Sect. 3, we gave insights about a new island gravitational fine-grained entropy proposal and more details will be given in future work.

References

1. A. Almheiri, R. Mahajan, J. Maldacena, Y. Zhao, J. High Energy Phys. **2020**, 024 (2020)
2. J.D. Bekenstein, Phys. Rev. D **7**, 2333 (1973)
3. S.W. Hawking, Commun. Math. Phys. **43**, 199 (1975)
4. S. Ryu, T. Takayanagi, Phys. Rev. Lett. **96**, 181602 (2006)
5. V.E. Hubeny, M. Rangamani, T. Takayanagi, J. High Energy Phys. **2007**, 062 (2007)
6. N. Engelhardt, A.C. Wall, J. High Energy Phys. **2015**, 1 (2015)
7. C. Akers, N. Engelhardt, G. Penington, M. Usatyuk, J. High Energy Phys. **2020**, 8 (2020)
8. N. Engelhardt, G. Penington, A. Shahbazi-Moghaddam, J. High Energy Phys. **2024**, 1 (2024)
9. A.C. Wall, Class. Quant. Grav. **31**, 225007 (2014)

Open Access This chapter is licensed under the terms of the Creative Commons Attribution 4.0 International License (http://creativecommons.org/licenses/by/4.0/), which permits use, sharing, adaptation, distribution and reproduction in any medium or format, as long as you give appropriate credit to the original author(s) and the source, provide a link to the Creative Commons license and indicate if changes were made.

The images or other third party material in this chapter are included in the chapter's Creative Commons license, unless indicated otherwise in a credit line to the material. If material is not included in the chapter's Creative Commons license and your intended use is not permitted by statutory regulation or exceeds the permitted use, you will need to obtain permission directly from the copyright holder.

A New Vertex Locator for the Next Generation LHCb Experiment

Enoch Ejopu

1 Introduction

The Large Hadron Collider beauty (LHCb) detector [1] is a single arm forward spectrometer dedicated to flavour physics measurements like the study of matter and antimatter states as seen in the precise determination of the B_s oscillation frequency [2], rare decays that only proceed via quantum-loop transitions and experience Cabibbo-Kobayashi-Maskawa (CKM) and helicity suppression [3], and other Standard Model (SM) precision tests. In addition, the experiment explores exotic hadrons as shown in the observation of a narrow pentaquark state, $P_c(4312)^+$ [4], electroweak physics and many other measurements in the forward region. All these will further the understanding of the Standard Model and test for potential extensions.

The detector has a unique coverage in pseudorapidity of $2 < \eta < 5$ with a 40 MHz full-detector readout rate and is currently being operated at an instantaneous luminosity of 2×10^{33} cm^{-2} s^{-1} with an expected integrated luminosity of 50 fb^{-1} by the end of LHC Run 4. The detector comprises the Vertex Locator (VELO) surrounding the particle collision point, the Upstream Tracker before and the Scintillating Fibre tracking stations after the dipole magnet. In addition, particle identification is facilitated by two Ring Imaging Cherenkov detectors, the electromagnetic and hadronic calorimeters, and the muon stations (see Fig. 1).

Enoch Ejopu on behalf of the LHCb VELO Group.

E. Ejopu (✉)
The University of Manchester, Manchester, UK
e-mail: enoch.ejopu@postgrad.manchester.ac.uk

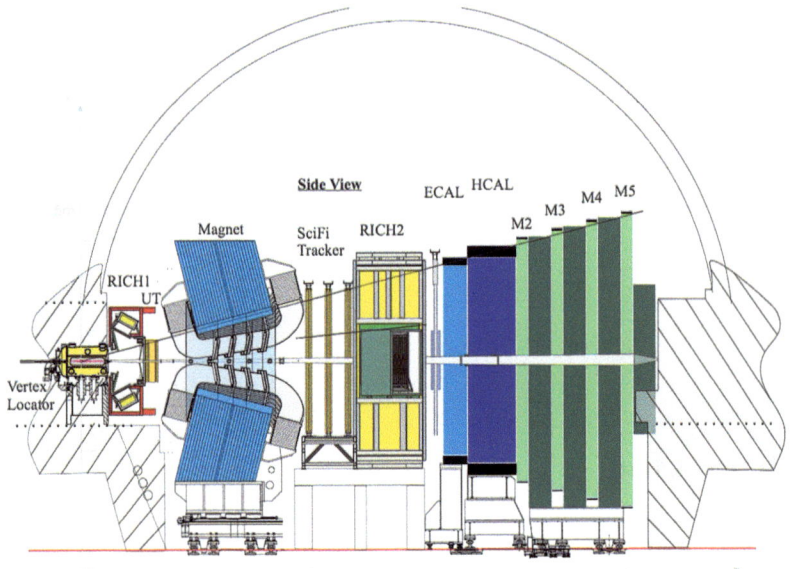

Fig. 1 A layout of the LHCb detector at CERN showing the VELO location. Reprinted under CC-BY-4.0 license from [5]. © 2024, CERN for the benefit of the LHCb collaboration

2 Vertex Locator

The VELO detector, designed for detecting charged particles near the proton-proton collision region, consists of 52 double-sided modules arranged in two halves of 26 modules each. The halves are housed in thin walled RF boxes that guide the beam wakefield and provide a secondary vacuum. Each module contains four hybrid planar pixel tiles, two of which are glued to each side of a microchannel cooling substrate. The VELO uses silicon pixel sensors, each pixel measuring $55\,\mu m \times 55\,\mu m$ in size. The sensors cover an approximately square area transverse to the beam axis with a central aperture in which the closest active area is 5.1 mm away from the beam axis. The VELO has a 2.8 Tbs^{-1} data rate and is built to withstand a maximum fluence of approximately 8×10^{15} 1 MeV n_{eq} cm^{-2}. The purpose of the detector is the reconstruction of trajectories of particles passing through it, which facilitates the detection and precision measurement of particles originating in vertices that are displaced from the primary pp collision region [6].

The sensors used must be operated in a low temperature environment, typically less than $-20\,°C$ for the entire life of the detector. However, the front-end ASICs operating in the detector vacuum constantly dissipate power and this must be removed by a cooling system. To do this efficiently, bi-phase CO_2 cooling was chosen. Here, the CO_2 flows in micro-channels within a silicon substrate, onto which the active components are glued. Presently, the detector is operational and has gathered data for the initial phase of the Run. Figure 2 shows the velo efficiency, evaluated using a downstream method, in bins of momentum in comparison of 2024 data and simulation.

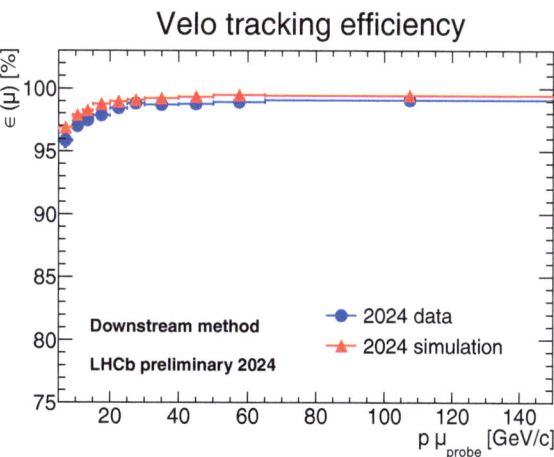

Fig. 2 Velo efficiency in bins of momentum, comparing 2024 data with simulation. Reprinted under CC-BY-4.0 license from [7]. © 2024, CERN for the benefit of the LHCb collaboration

3 LHCb Upgrade II

To achieve ultimate sensitivity in flavour physics, and thereby increase the potential to discover new phenomena in physics, the available dataset has to be vastly increased. This requires an upgraded detector to operate with a tenfold increased instantaneous luminosity, thus increasing the integrated luminosity to approximately 300 fb^{-1} [8] by the end of High-Luminosity LHC (HL-LHC) era. This requires the addition of precise timing capabilities to several sub-detectors to disentangle different collisions and thereby facilitate the required reconstruction precision.

For the upgraded VELO, this necessitates a novel detector that satisfies the required position and timing resolution in addition to the radiation tolerance. Candidate silicon sensors like 3D sensors [9], thin planar sensors [10], Silicon Electron Multiplier sensors [11] and Low Gain Avalanche Detectors (LGADs) [12] technologies are currently under study. To achieve the performance requirements, sensors need a time resolution as good as 30 picoseconds per track in order to assign each particle to the correct vertex [13]. One of the candidates are LGADs, which are under study by the author.

One particular method that has been used to characterise the performance of LGAD sensors is the Two Photon Absorption (TPA) technique [14]. Here, a point-like laser probe formed by energy confinement and known as the "voxel" is used to induce electron-hole pairs in a precisely defined region. This facilitates the study of the sensor depletion voltage, electric field, and timing resolution. To obtain a picture of the sensor response over the entire pixel cell, the voxel is scanned in all three dimensions. The set-up has been fully installed and initial time resolutions have been determined. Reflection of the laser on the metallisation of the back plane has been observed. Studies in the coming years will establish which of the technologies best satisfies the requirements for the VELO detector for LHCb Upgrade II.

Acknowledgments Many Thanks to the Commonwealth Scholarship Commission in the UK for funding my PhD studies.

References

1. The LHCb Collaboration, LHCb Detector Performance. Int. J. Mod. Phys. A **30**, 1530022 (2015)
2. LHCb Collaboration, Precise determination of the B_s^0–$\overline{B_s^0}$ oscillation frequency. Nat. Phys. **18**, 1–5 (2022). https://doi.org/10.1038/s41567-021-01394-x
3. LHCb Collaboration, Analysis of neutral B-meson decays into two muons. Phys. Rev. Lett. **128**, 041801 (2022). https://doi.org/10.1103/PhysRevLett.128.041801
4. LHCb Collaboration, Observation of a narrow pentaquark state, $P_c(4312)^+$, and of two-peak structure of the $P_c(4450)^+$. Phys. Rev. Lett. **122**, 222001 (2019)
5. M. Mulder, C.J.G. Onderwater, LHCb Collaboration, The LHCb upgrade I. J. Instrum. **19**, P05065 (2024)
6. LHCb Collaboration, The LHCb Upgrade I. LHCB-DP-2022-002 (2023)
7. LHCb Collaboration, Track reconstruction efficiency mass plots and efficiency plots for 2024 data and simulation, CERN Document Server, Record 2913624 (2024)
8. LHCb Collaboration, Physics case for an LHCb Upgrade II - Opportunities in flavour physics, and beyond, in the HL-LHC era. LHCB-PUB-2018-009, CERN-LHCC-2018-027 (2018)
9. A. Lai, A system approach towards future trackers at high luminosity colliders: the TIMESPOT project, in *IEEE NSS/MIC* (2018), pp. 1–3
10. J. Duarte-Campderros, Pixelated 3D sensors for tracking in radiation harsh environments. JPS Conf. Proc. **34**, 010005 (2021)
11. M.M. Halvorsen, V. Coco, E.L. Gkougkousis, P. Collins, O. Girard, The silicon electron multiplier sensor. Nucl. Instrum. Meth. A **1041**, 167325 (2022)
12. Pellegrini et al., Technology developments and first measurements of Low Gain Avalanche Detectors (LGAD) for high energy physics applications. Nucl. Instrum. Meth. A **765**, 12–16 (2014)
13. Casado et al., A high-granularity timing detector for the ATLAS Phase-II upgrade. Nucl. Instrum. Meth. A **1032**, 166628 (2022)
14. M. Rumi, J.W. Perry, Two-photon absorption: an overview of measurements and principles. Adv. Opt. Photon. **2**, 451–518 (2010)

Open Access This chapter is licensed under the terms of the Creative Commons Attribution 4.0 International License (http://creativecommons.org/licenses/by/4.0/), which permits use, sharing, adaptation, distribution and reproduction in any medium or format, as long as you give appropriate credit to the original author(s) and the source, provide a link to the Creative Commons license and indicate if changes were made.

The images or other third party material in this chapter are included in the chapter's Creative Commons license, unless indicated otherwise in a credit line to the material. If material is not included in the chapter's Creative Commons license and your intended use is not permitted by statutory regulation or exceeds the permitted use, you will need to obtain permission directly from the copyright holder.

The Compact Linear Collider: Physics Potential

Jan Klamka

1 Introduction

The CLIC implementation plan assumes three running stages: first with $\sqrt{s} = 380\,\text{GeV}$, and two subsequent high-energy stages at 1.5 and 3 TeV [1]. It is assumed that integrated luminosities of $1\,\text{ab}^{-1}$, $2.5\,\text{ab}^{-1}$ and $5\,\text{ab}^{-1}$ will be collected, respectively, including $100\,\text{fb}^{-1}$ for an energy scan around the top quark pair-production threshold at the first stage. The high luminosity and wide energy range enable a broad physics programme, while staged implementation allows this to be adjusted in case of potential discoveries.

The studies of the CLIC physics potential rely on state-of-the-art software. The main event generator is WHIZARD, and most of the results are based on detailed GEANT4 simulation, most recently implemented in DD4HEP. Reconstruction frameworks include Conformal Tracking and VLC jet clustering algorithm dedicated for CLIC, as well as PandoraPFA for particle flow, algorithms for flavour tagging, isolation, and more. These tools build on the ILCSoft framework and are now implemented in KEY4HEP [2]. Beam spectra and beam-induced backgrounds are also included in the full simulation, with dedicated timing cuts applied at the reconstruction level to reduce a large background of beam-induced $\gamma\gamma \to$ hadrons events. Implementation of CLICdet in DELPHES also enables fast simulation studies.

Jan Klamka on behalf of the CLICdp Collaboration.

J. Klamka (✉)
Faculty of Physics, University of Warsaw, Warsaw, Poland
e-mail: jan.klamka@fuw.edu.pl

2 Higgs Physics

At $\sqrt{s} = 380$ GeV the Higgs-strahlung channel ($e^+e^- \to$ ZH) allows the identification of Higgs production regardless of the H^0 decay mode, by using the Z $\to e^+e^-, \mu^+\mu^-$ recoil mass [3]. Therefore, it is possible to measure g_{HZZ} coupling and BR(H $\to$ invisible) in a model-independent way, the latter of which can be constrained to 1% at 95% C.L. [4]. Higgs-strahlung can also be used to measure the branching ratios for most of the Higgs decays. A global fit to $\sigma \times$ BR measured in the HZ and VBF ($e^+e^- \to H\nu\nu$) channels provides an estimate of most of the Higgs couplings with a precision of less than 1%, without any assumptions on physics models [3, 5].

The Higgs self-coupling is connected to multiple open problems in fundamental physics. Direct access to the g_{HHH} coupling is possible only above $\sqrt{s} \sim 500$ GeV. Therefore, the high-energy stages of CLIC are well suited for the extraction of g_{HHH} in the ZHH (at 1.5 TeV) and HH$\nu\nu$ (at 3 TeV) production channels. Based on the measured HH distributions, it is expected that g_{HHH} will be measured with an uncertainty of $^{+11}_{-8}$% [6]. CLIC is the future Higgs factory that will give the earliest measurement of g_{HHH} with $\mathcal{O}(10\%)$ precision [7].

3 Top Quark Physics

CLIC is also well suited for top-quark studies in multiple production channels. The energy of the first CLIC stage is just above the maximum for the pair-production cross section while the second stage, with $\sqrt{s} = 1.5$ TeV, is close to the maximum rate for $t\bar{t}$H production (the luminosity of a linear collider increases with energy), with 3 TeV data allowing a direct measurement of the top Yukawa coupling. An interesting physics case is also provided by $t\bar{t}$ production in VBF, which is very sensitive to BSM effects in many scenarios [8].

The top-quark mass can be measured at CLIC using three different methods [8]. The most precise involves a dedicated scan of the $t\bar{t}$ production threshold, with 10 energy points and 10 fb^{-1} collected at each energy. The cross section shape in this region is sensitive to the top-quark mass and width, which can be extracted using a template fit. With the main systematic uncertainties coming from the strong coupling constant α_s and the top Yukawa coupling, 20 MeV statistical and 50 MeV total uncertainty on the top mass can be achieved. The second method is based on the measurement of the $t\bar{t}\gamma$ events above the threshold, with a reconstructed hard ISR photon, in which the mass is extracted from the cross section dependence on the effective centre-of-mass energy spectrum, similar to the threshold scan. The precision of this method is around 140 MeV. The top mass can also be directly reconstructed from the invariant mass distribution; however, this method is limited by large theoretical and jet energy scale uncertainties. Further top-sector measurements include electroweak couplings, CP properties and searches for FCNC and compositeness, among others.

4 Beyond Standard Model Searches

Direct and indirect searches for new physics are high priorities in the CLIC programme. The effective field theory framework provides tools that are sensitive to high new-physics scales in a model-independent way by using precision observables. Based on measurements of Higgs and top-quark couplings, WW production and 2-fermion processes at CLIC, it is possible to test scales in the 10–100 TeV range, depending on the operator [9]. This extends the reach of the HL-LHC by orders of magnitude. On the other hand, new scalar singlets could be probed for mixings $\sin^2\gamma > 0.24\%$ with the SM Higgs and the Higgs compositeness tested up to the scale of 10 TeV. Supersymmetric particles could be discovered in the vast majority of the parameter space up to masses of 1.5 TeV, and a dark photon up to its mass of 20 TeV. Here, and in many other direct searches, CLIC can also surpass the HL-LHC sensitivity, probing many corners of the parameter space and mass ranges almost up to the kinematic limit [10].

5 Conclusion

The clean environment at CLIC, its high collision energies and electron beam polarisation enable unprecedented precision in Higgs, electroweak and top quark studies. Key highlights of the programme are a determination of all Higgs couplings, studies of CP violation effects and a top-threshold scan. At high energy stages, CLIC is well-suited for both direct and indirect BSM physics searches. Physics studies show that CLIC surpasses the HL-LHC in its potential for precision measurements and is competitive in the exploration of many new physics scenarios.

References

1. P. Burrows et al. (eds.), *The Compact Linear Collider (CLIC) - 2018 Summary Report*, CERN Yellow Reports: Monographs, vol. 1802 (2018). https://doi.org/10.23731/CYRM-2018-002
2. G. Ganis et al., Eur. Phys. J. Plus **137**(1), 149 (2022). https://doi.org/10.1140/epjp/s13360-021-02213-1
3. H. Abramowicz et al., Eur. Phys. J. C **77**(7), 475 (2017). https://doi.org/10.1140/epjc/s10052-017-4968-5
4. K. Mękała et al., Eur. Phys. J. Plus **136**(2), 160 (2021). https://doi.org/10.1140/epjp/s13360-021-01116-5
5. A. Robson, P. Roloff, Updated CLIC luminosity staging baseline and Higgs coupling prospects (2018). Preprint at https://arxiv.org/pdf/1812.01644
6. P. Roloff et al., Eur. Phys. J. C **80**(11), 1010 (2020). https://doi.org/10.1140/epjc/s10052-020-08567-7
7. J. de Blas et al., J. High Energy Phys. **01**, 139 (2020). https://doi.org/10.1007/JHEP01(2020)139

8. H. Abramowicz et al., J. High Energy Phys. **11**, 003 (2019). https://doi.org/10.1007/JHEP11(2019)003
9. J. de Blas et al., *The CLIC Potential for New Physics*, CERN Yellow Reports: Monographs, vol. 3/2018 (2018). https://doi.org/10.23731/CYRM-2018-003
10. P. Roloff et al., The compact linear e^+e^- collider (CLIC): physics potential (2018). Preprint at https://arxiv.org/pdf/1812.07986

Open Access This chapter is licensed under the terms of the Creative Commons Attribution 4.0 International License (http://creativecommons.org/licenses/by/4.0/), which permits use, sharing, adaptation, distribution and reproduction in any medium or format, as long as you give appropriate credit to the original author(s) and the source, provide a link to the Creative Commons license and indicate if changes were made.

The images or other third party material in this chapter are included in the chapter's Creative Commons license, unless indicated otherwise in a credit line to the material. If material is not included in the chapter's Creative Commons license and your intended use is not permitted by statutory regulation or exceeds the permitted use, you will need to obtain permission directly from the copyright holder.

The Upgrade of the T2K Near Detector

Nataliya Skrobova

1 Introduction

The T2K (Tokai-to-Kamioka) experiment is a world leading long baseline oscillation experiment. It discovered non-zero θ_{13} mixing angle with accelerator neutrinos, obtained CP violation hints in neutrino sector [1], and has leading sensitivities in Δm^2_{23} and θ_{23} octant. An off-axis high-intensity neutrino beam is produced in J-PARK, the near detector complex measures the beam parameters, and then oscillated neutrino beam is measured in the far detector Super-Kamiokande. The ND280 is an of-axis near detector. Three time-projection chambers (TPCs) and two Fine-Grained Detectors (FGDs) form a tracking system which together with a π_0 detector (P0D) is covered by electromagnetic calorimeters, and a solenoid coil. Side Muon Range Detectors (SMRD) are embedded in the iron yokes of the magnet. Magnetic field is 0.2 T. The schematic view of ND280 is shown in Fig. 1 (left). The role of ND280 is to tune and constrain the flux and cross-section systematics model by selecting neutrino interactions before the oscillations. The impact on T2K oscillation analysis is dramatic: systematic uncertainties are reduced from 15–20% to ∼5%. Despite a significant success in systematic uncertainties reduction, the initial ND280 design has its limitations. ND280 has a non-isotropic detection efficiency in comparison to 4π sensitivity in the Super-Kamiokande detector. The detection threshold for protons is about 450 MeV/c (∼100 MeV E_{kin}) which is relatively high. Moreover

---

Nataliya Skrobova for the T2K collaboration.

N. Skrobova (✉)
INR, Institute for Nuclear Research, Moscow, Russia

LPI, Lebedev Physics Institute, Moscow, Russia
e-mail: skrobova@lebedev.ru

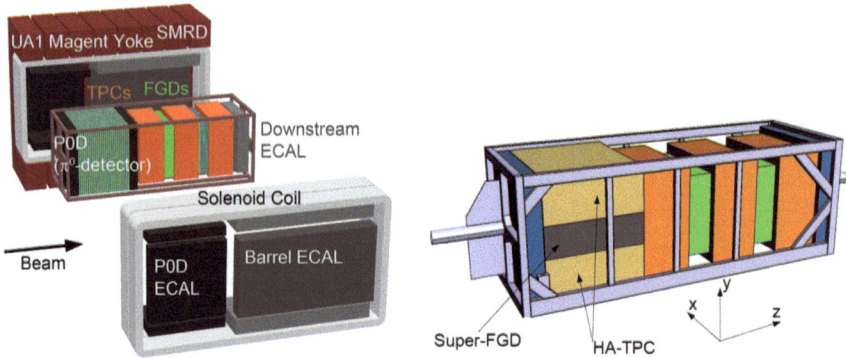

Fig. 1 The schematic view of the initial Near Detector complex (left) and the proposed upgraded configuration (right) with the SuperFGD and HA-TPCs. The ECal, solenoid coil, magnet yoke and SMRD will be kept at place but are not shown as well as the ToF planes

ND280 is not capable to detect neutrons. Hadronic information is essential to improve neutrino interaction modelling and hence the energy reconstruction for the oscillation analysis.

The upgrade of ND280 [2] is intended to handle these limitations and to substantially improve systematics. The P0D detector is replaced by a setup of detectors: a new scintillator target (SuperFGD), two High-Angle TPCs (HA-TPCs) and a Time-of-Flight (ToF) detector. See Fig. 1 (right).

2 New Detectors

2.1 The SuperFGD Detector

The SuperFGD (Super Fine-Grained Detector) is a fully active and highly granular scintillator neutrino detector. While FGDs consist of long scintillation strips, which do not allow reconstruction of high-angle tracks, SuperFGD consists of ~2 million scintillation cubes, which allow reconstruction of particles going in all directions. Polysterene cubes are injection-molded, doped with PTP (1.5%) and POPOP (0.01%) and have reflector coating. Each cube has tree drilled holes going in orthogonal projections. Scintillation light is collected by 1 mm Kuraray Y-11(200) MS wave-length shifting (WLS) fibers. Silicon photo-multipliers (SiPMs) are placed at one side of each WLS fiber and are used to read signals (model: S13360-1325PE Hamamatsu MPPC). A free side of each fiber is connected to a LED calibration system. Multiple beam tests of prototypes have been performed to characterize the SuperFGD performance [3, 4]. The $(8 \times 24 \times 48\ cm^3)$ prototype was tested at the beam in CERN [5]. The raw light yield was 53.7 p.e. (51.0 p.e.) per MIP for events recorded on channels with the 8 cm (24 cm) fibers. For the long

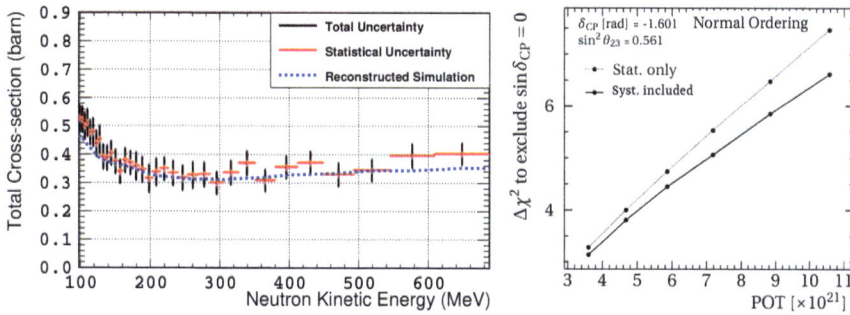

Fig. 2 Left: The total neutron-CH cross section as a function of neutron kinetic energy. The black vertical bars represent the total uncertainty and the red the statistical uncertainty. The Geant4 Bertini model is shown in blue. Right: Sensitivity to Charge-Parity violation as a function of Protons-on-Target when considering only statistical uncertainties (dotted) and when using the systematic uncertainties with new detectors constraints (solid)

fibers which were parallel to the beam direction energy released in several cubes contributes to a signal in one channel making it difficult to determine the light yield. Optical crosstalk was found to be 2.94% per cube side. The dE/dx resolution for a single cube is 25% and improves up to 10% for longer tracks. It was demonstrated that one channel time resolution is better than 1 ns, further improved with higher amplitudes and larger number of channels [6].

Two SuperFGD prototypes ($8 \times 24 \times 48$ cm^3 and $8 \times 8 \times 32$ cm^3) were exposed to a neutron beam at LANL. This is an important measurement to demonstrate a neutron detection capability in an energy range relevant for T2K [7]. Neutron energy is reconstructed using time-of-flight information between the arrival of gammas and neutron secondary interaction. For such type of measurements time resolution is a crucial factor. It was tested that like in case of muons averaging information from different channels allows to improve time resolution by a factor of $\sqrt{N}$ where N is a number of channels. For selected neutron events with at least 3 cubes intrinsic time resolution is about 0.5 ns (using time measurements from fibers orthogonal to the beam direction). The improvement in the neutrino flux constraints due to the ability of neutron energy reconstruction is discussed in [8] In the presence of neutron-nucleus interactions, the signal event rate decreases exponentially along the coordinate parallel to the beam. This allows to measure the neutron cross-section as a function of its kinetic energy (see Fig. 2, left).

2.2 High-Angle Time Projection Chambers

Two new additional "horizontal" TPCs will surround the SuperFGD detector from the top and bottom sides. Their configuration allows the identification and momentum reconstruction for up and downward going tracks with respect to the neutrino beam, which is complementary to the forward-going particle information from the current, vertical TPCs. The design of HA-TPCs is similar to the existing ND280 TPCs. The main differences to the existing TPCs are the replacement of the standard bulk-MicroMegas with new resistive bulk-MicroMegas and a new field cage design to minimize dead space and maximize the tracking volume. Encapsulated Resistive Anode Micromegas (ERAM) with grounded mesh and anode at a positive amplification voltage provide several advantages. Since the charge is spread over several pads the spatial resolution is improved by a factor of $\sqrt{12}$. This allows to use larger pads for the same spatial resolution as for the bulk Micromegas detectors. Insulation of the resistive anode from the pads suppresses formation of sparks and limits their intensity. Therefore, it is possible to get rid of the cumbersome anti-spark protection circuitry. Another advantage of the anode encapsulation is that the detection plane is fully equipotential. This leads to more uniform field, less track distortions, and module flexibility.

Characterization of the charge spreading (RC), gain of ERAMs using X-rays, measurements of the spatial and dE/dx resolution were performed in several test beam campaigns at CERN and DESY: [9–11]. Achieved spatial resolution is better than 800 μm and the dE/dx resolution is better than 10% for all track angles and drift distances of interest for T2K.

2.3 Time-of-Flight Detectors

Time-of-Flight system consists of 6 planes of 2.3×2.5 m^2 size, each plane is made of 20 plastic scintillator bars. These planes fully envelope the SuperFGD and High-Angle TPCs. Each scintillator bar is read out on both ends with an array of 8 SiPMs. An average time resolution of about 0.14 ns is achieved for a single bar when measured with cosmic muons [12]. This result is very important to determine the direction of particles and the vertex time, and therefore to suppress backgrounds.

3 Summary

The main goal of the ND280 upgrade is to reduce systematic uncertainties associated with neutrino flux and cross-section modeling for future studies of neutrino oscillations. The upgraded ND280 detector will be able to perform a full exclusive reconstruction of the final state from neutrino-nucleus interactions, including measurements of low momentum protons (~50 MeV E_{kin}), pions and for the first time, event-by event measurements of neutron kinematics. The physics sensitivity that can be expected from the upgraded detector is described in [13]. The expected sensitivity to Charge-Parity violation searches for the level of systematics expected after the ND280 upgrade is shown in Fig. 2, right.

Acknowledgments This work was supported by the RSF grant 22-72-00054.

References

1. K. Abe, et al., Measurements of neutrino oscillation parameters from the T2K experiment using 3.6×10^{21} protons on target. Eur. Phys. J. C **83**, 782 (2023). https://doi.org/10.1140/epjc/s10052-023-11819-x
2. K. Abe, et al., T2K ND280 Upgrade – Technical Design Report. arXiv:1901.03750 [hep-ex]. https://arxiv.org/abs/1901.03750
3. A. Blondel et al., A fully-active fine-grained detector with three readout views. J. Instrum. **13**, P02006 (2018). https://doi.org/10.1088/1748-0221/13/02/P02006
4. O. Mineev et al., Beam test results of 3D fine-grained scintillator detector prototype for a T2K ND280 neutrino active target. Nucl. Instrum. Meth. A **923**, 134–138 (2019). https://doi.org/10.1016/j.nima.2019.01.080
5. A. Blondel et al., The SuperFGD Prototype charged particle beam tests. J. Instrum. **15**, P12003 (2020). https://doi.org/10.1088/1748-0221/15/12/P12003
6. I. Alekseev et al., SuperFGD prototype time resolution studies. J. Instrum. **18**, P01012 (2023). https://doi.org/10.1088/1748-0221/18/01/P01012
7. A. Agarwal et al., Total neutron cross-section measurement on CH with a novel 3D-projection scintillator detector. Phys. Lett. B **840**, 137843 (2023). https://doi.org/10.1016/j.physletb.2023.137843
8. S. Gwon et al., Neutron detection and application with a novel 3D-projection scintillator tracker in the future long-baseline neutrino oscillation experiments. Phys. Rev. D **107**, 032012 (2023). https://doi.org/10.1103/PhysRevD.107.032012
9. D. Attié et al., Performances of a resistive Micromegas module for the time projection chambers of the T2K near detector upgrade. Nucl. Instrum. Meth. A **957**, 163286 (2020). https://doi.org/10.1016/j.nima.2019.163286
10. D. Attié et al., Characterization of resistive micromegas detectors for the upgrade of the T2K near detector time projection chambers. Nucl. Instrum. Meth. A **1025**, 166109 (2022). https://doi.org/10.1016/j.nima.2021.166109
11. D. Attié et al., Analysis of test beam data taken with a prototype of TPC with resistive micromegas for the T2K near detector upgrade. Nucl. Instrum. Meth. A **1052**, 168248 (2023). https://doi.org/10.1016/j.nima.2023.168248
12. A. Korzenev et al., A 4π time-of-flight detector for the ND280/T2K upgrade. J. Instrum. **17**, P01016 (2022). https://doi.org/10.1088/1748-0221/17/01/P01016
13. S. Dolan et al., Sensitivity of the upgraded T2K near detector to constrain neutrino and antineutrino interactions with no mesons in the final state by exploiting nucleon-lepton correlations. Phys. Rev. D **105**, 032010 (2022). https://doi.org/10.1103/PhysRevD.105.032010

Open Access This chapter is licensed under the terms of the Creative Commons Attribution 4.0 International License (http://creativecommons.org/licenses/by/4.0/), which permits use, sharing, adaptation, distribution and reproduction in any medium or format, as long as you give appropriate credit to the original author(s) and the source, provide a link to the Creative Commons license and indicate if changes were made.

The images or other third party material in this chapter are included in the chapter's Creative Commons license, unless indicated otherwise in a credit line to the material. If material is not included in the chapter's Creative Commons license and your intended use is not permitted by statutory regulation or exceeds the permitted use, you will need to obtain permission directly from the copyright holder.

Hyper-Kamiokande: Physics Potential, Calibration and Detector Systematics

Sam J. Jenkins

1 Introduction

The upcoming Hyper-Kamiokande [1] experiment is a next-generation water Cherenkov detector, which will be situated in Gifu-ken, Japan. Hyper-K will have a broad physics program, ranging from precise measurements of the CP violation phase δ_{CP}, to astrophysical sources such as solar, atmospheric and supernova neutrinos, to proton decay.

2 Experimental Setup

2.1 Hyper-Kamiokande Detector

The Hyper-K detector will consist of a cylindrical tank measuring 71 m in height and 68 m in diameter, filled with ultra-pure water (Fig. 1). This equates to a total (fiducial) volume of 258 kton (187 kton). The detector will be situated in the Tochibora mine 8 km south of the Super-K site, under the peak of the Nijyuugo-yama mountain, resulting in an overburden of 650 m (1750 m.w.e).

On behalf of the Hyper-Kamiokande collaboration.

S. J. Jenkins (✉)
University of Liverpool, Liverpool, UK
e-mail: s.j.jenkins@liverpool.ac.uk

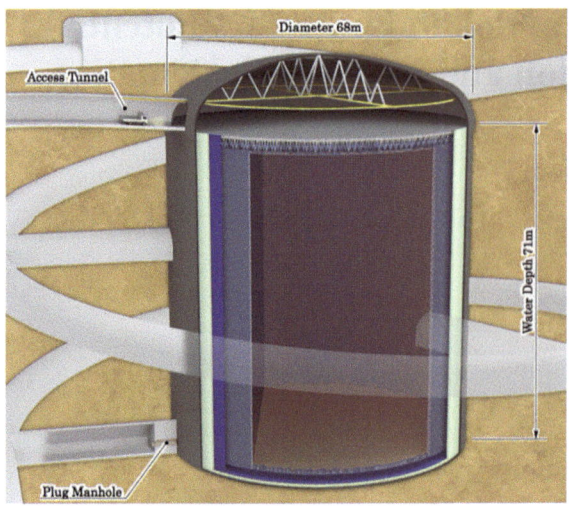

Fig. 1 Schematic of the Hyper-K detector

Fig. 2 mPMT viewed from above

The detector volume will be divided into inner (ID) and outer (OD) detector regions, optically separated by high reflectivity Tyvek sheets. The ID will be instrumented by 20,000 50 cm photomultiplier tubes (PMTs), corresponding to a photo-coverage of 20%. In addition, approximately 800 multi-PMTs (mPMTs) will be evenly dispersed; these are 19 PMTs (8 cm) housed in a 50 cm diameter pressure-sealed vessel (Fig. 2), providing increased spatial and timing resolution. The OD, which acts as a veto region, will contain 3600 PMTs (8 cm). Each of these PMTs will sit at the centre of a 30 cm square wavelength shifting (WLS) plate, in order to increase the photo-coverage in this region.

2.2 J-PARC Neutrino Beam

Hyper-K will act as the far detector for a long-baseline neutrino oscillation experiment using an upgraded version of the J-PARC neutrino beam, which produces an almost pure muon (anti)neutrino beam. This is directed 2.5° off-axis from the far detector, resulting in a narrow band beam with a peak E_ν of around 600 MeV. At a distance of 295 km, this coincides with the first oscillation maximum.

As of 2021, the neutrino beam was operating at 515 kW. Upgrades to the magnet power supplies have now increased the repetition rate from 2.48 to 1.36 s per cycle, achieving the original design of 750 kW. Towards the operation of Hyper-K, further upgrades including a boost of the repetition rate to 1.16 s per cycle are planned to bring the beam power to 1.3 MW [2].

2.3 Near and Intermediate Detectors

The ND280 detector sits 280 m downstream of the beam production point, at the same off-axis angle as the far detector, and is comprised of multiple different subdetectors. To improve phase space acceptance and tracking thresholds, ND280 has undergone a recent upgrade. Full details on ND280 and the upgrade can be found in [3] and [4], respectively. The near detector complex also features a second detector, INGRID [5], a layered iron-scintillator detector which covers a range of off-axis angles, from 0° to 1.1°. Measurements of neutrino flux at a range of energies, wrong sign backgrounds, and cross sections on different targets are performed to constrain systematic uncertainties on oscillation measurements.

To complement these detectors, the Intermediate Water Cherenkov Detector (IWCD) is planned, with a 1 km baseline. This 500 ton detector will be instrumented with 400 mPMTs and positioned in a vertical shaft, allowing it to view the neutrino beam at between 1° and 4° off-axis. Along with measuring flux at various off-axis angles, IWCD will make high-purity ν_e and $\bar{\nu}_e$ cross section measurements from the intrinsic contamination in the beam.

3 Physics Goals

Hyper-K will cover a wide range of physics, with just a few topics covered here. Further details can be found in [1].

3.1 Oscillation Physics

One of the primary physics goals of Hyper-K is searching for CP violation in neutrino oscillations. Electron (anti)neutrino appearance in the muon (anti)neutrino

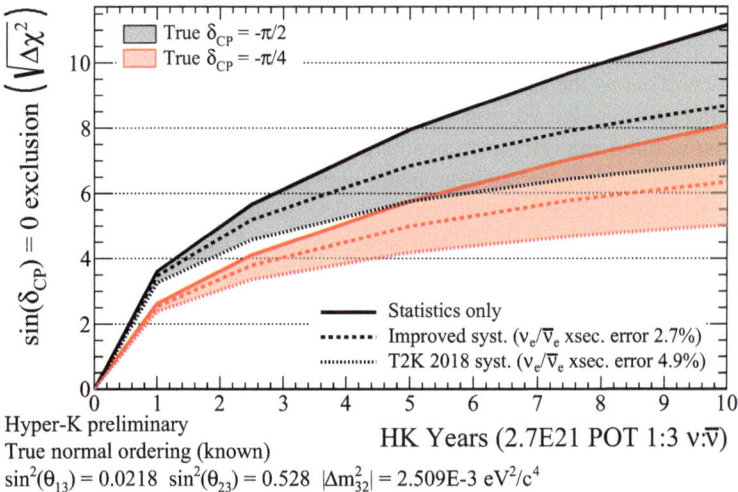

Fig. 3 CP conservation exclusion as a function of operating time, for true $\delta_{CP} = -\pi/2$ and $-\pi/4$. The mass hierarchy is assumed to be normal

beam is observed; any difference between appearance rates of neutrinos and antineutrinos implies violation of CP symmetry. While current experiments are statistically limited, after 10 years of operation Hyper-K expects to see over 1000 ν_e and $\bar{\nu}_e$ events, and as such will be limited by systematic uncertainties.

Figure 3 shows the ability to exclude CP conservation as a function of time for true values of δ_{CP}. Exclusion depending on systematic uncertainty levels is shown, where the improved systematics have a reduction in the $\nu_e/\bar{\nu}_e$ cross section uncertainty. This shows the importance of understanding these systematics and the role of IWCD in ν_e cross section measurements; CP conservation can be excluded years earlier with improved systematics. After 10 years, CP conservation can be excluded in 60% of δ_{CP} space at 5σ, assuming improved systematics.

Due to the relatively short baseline, accelerator neutrinos at Hyper-K do not have high sensitivity to matter effects. However, atmospheric (anti)neutrinos travelling through the Earth will have their oscillations enhanced (suppressed) if the mass hierarchy is normal, where the reverse is true for an inverse hierarchy. This allows for determination of the hierarchy via upward-going atmospherics. In addition, combining beam and atmospheric samples helps to break degeneracies between parameters, improving sensitivities to both δ_{CP} and the mass hierarchy.

3.2 Proton Decay

While not allowed by the standard model, proton decay is predicted by Grand Unified Theories. With its large mass of protons, Hyper-K can produce world leading measurements of this process. One favoured decay mode Hyper-K can observe is the

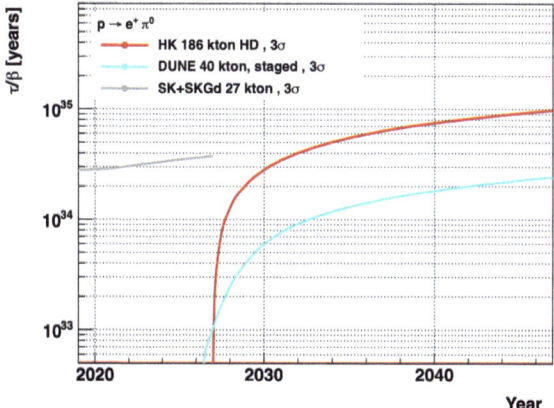

Fig. 4 3σ discovery potential for $p \to e^+\pi^0$ as a function of years

$p \to e^+\pi^0$ channel, which produces three back-to-back electromagnetic showers. The 3σ sensitivity to this process is shown in Fig. 4. It is clear that Hyper-K is the only realistic chance of probing proton lifetimes of up to 10^{35} years.

4 Systematic Requirements and Approaches

It is clear that constraining systematic errors is vital for Hyper-K to make precision measurements of oscillation parameters. However, systematic effects are also important as they can be degenerate with the parameters to be measured. Figure 5 shows how a $12°$ shift in δ_{CP} can be mimicked by a 0.5% shift in the detector energy scale. In order to constrain systematic uncertainties, a diverse program of calibration sources is in preparation. Light injection systems will be used to illuminate PMTs with known quantities of light on a daily basis to measure water parameters and PMT response, whilst an electron LINAC and multiple radioactive sources will be deployed at less regular intervals for dedicated detector calibration runs. In addition, a number of the ID PMTs will be calibrated before installation into the tank, providing references for surrounding PMTs. Further details of the calibration program are given in [1].

5 Current Status

Excavation of the Hyper-K site is on schedule, with the full cavern nearing completion. PMT production and quality assurance are underway, toward the experiment beginning to take data in 2028.

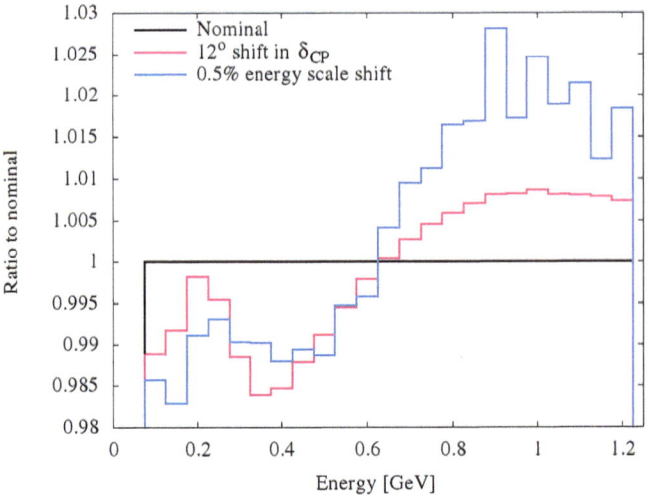

Fig. 5 Variation of the 1-ring electron-like sample for a 12° shift in δ_{CP} (red) and 0.5% in energy scale (blue). Reprinted under CC BY-NC-ND 4.0 from [6]. ©2021, The Author(s)

References

1. K. Abe et al., (Hyper-Kamiokande Proto-Collaboration), Hyper-Kamiokande design report. arXiv:1805.04163v2 [physics.ins-det] (2018). https://doi.org/10.48550/arXiv.1805.04163
2. T. Yasui, Y. Sato, H. Hoichi, S. Igarashi, J-PARC MR upgrade commissioning status and future plan, in *NuFACT 2023*, Seoul, Korea (2023)
3. K. Abe et al., (T2K Collaboration), The T2K experiment. Nucl. Instrum. Methods Phys. Res. Sect. A **659**(1), 106–135 (2011). https://doi.org/10.1016/j.nima.2011.06.067
4. Lux, T., The upgrade of the T2K ND280 detector. J. Phys. Conf. Ser. **2374**, 012036 (2022). https://doi.org/10.1088/1742-6596/2374/1/012036
5. K. Abe et al., Measurements of the T2K neutrino beam properties using the INGRID on-axis near detector. Nucl. Instrum. Methods Phys. Res. Sect. A **694**, 211–223 (2012). https://doi.org/10.1016/j.nima.2012.03.023
6. L. Munteanu, Long-baseline neutrino oscillation sensitivities with Hyper-Kamiokande, in *PoS (NuFact2021)*, 056 (2022). https://doi.org/10.22323/1.402.0056

Open Access This chapter is licensed under the terms of the Creative Commons Attribution 4.0 International License (http://creativecommons.org/licenses/by/4.0/), which permits use, sharing, adaptation, distribution and reproduction in any medium or format, as long as you give appropriate credit to the original author(s) and the source, provide a link to the Creative Commons license and indicate if changes were made.

The images or other third party material in this chapter are included in the chapter's Creative Commons license, unless indicated otherwise in a credit line to the material. If material is not included in the chapter's Creative Commons license and your intended use is not permitted by statutory regulation or exceeds the permitted use, you will need to obtain permission directly from the copyright holder.

Highlights of Recent Standard Model Measurements with ATLAS

Daniel Lewis

1 Introduction

The Standard Model (SM) of particle physics describes the fundamental constituents of matter and their interactions. As the search continues at the LHC for indicators of new physics, the SM is subjected to advancing levels of scrutiny. Recent results from the ATLAS experiment [1] take advantage of the full Run-1 and Run-2 LHC datasets to provide precise tests of the SM, and also to explore new regions and aspects. The first results using the Run-3 dataset at a new, higher centre-of-mass energy of 13.6 TeV, are also now becoming available. This work summarises a selection of recent results.

2 Measurements of Single Bosons

The study of the individual electroweak bosons of the SM provide powerful tests due to their large production cross-sections and clean detector signatures. An extremely precise measurement of the Z boson transverse momentum (p_T) spectrum at a centre-of-mass energy of 8 TeV [2], is used to test predictions of up to next-to^3-leading order (N^{3}LO) in perturbative QCD. The Sudakov (low-p_T) region of the spectrum is only able to be described well when also including approximate next-

---

Daniel Lewis on behalf of the ATLAS Collaboration.

D. Lewis (✉)
LAPP, Annecy, France
e-mail: daniel.lewis@cern.ch

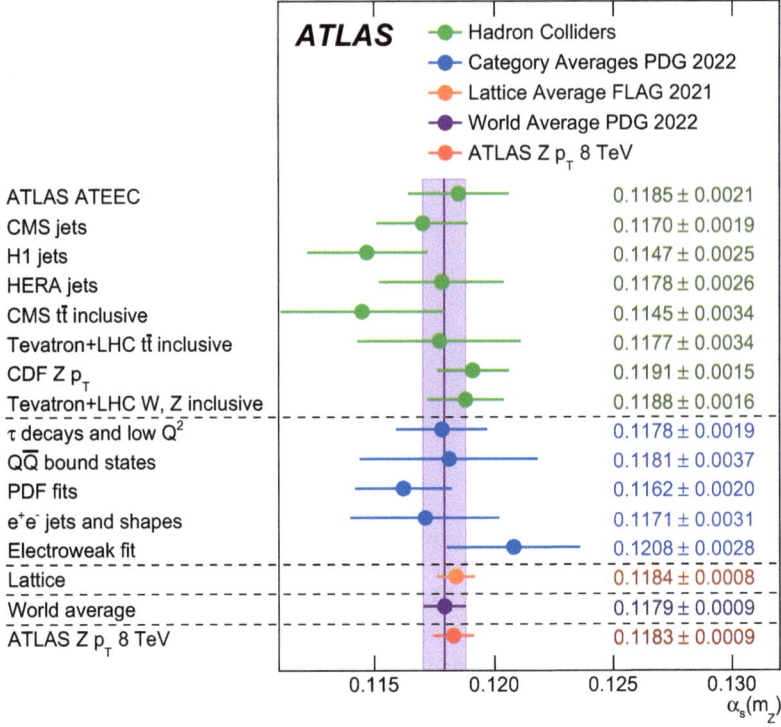

Fig. 1 The extracted value of $\alpha_s(m_Z)$ from the Z boson p_T spectrum at 8 TeV, compared to the world average and other experimental determinations. Reprinted under CC-BY-4.0 license from [3]. © 2024, CERN for the benefit of the ATLAS Collaboration

to^4-leading logarithmic (N^{4}LL) resummed contributions. The measurements are performed in the full decay phase space of the leptons, which is achieved via a decomposition of the cross-section into nine spherical harmonics described by angular variables of the leptons.

These cross-section measurements are also used as an input to a fit to extract an experimental determination of the strong coupling constant at the scale of the Z mass, $\alpha_s(m_Z)$ [3]. This fit benefits greatly from the reduction of modelling uncertainties due to the full lepton decay phase space, and the fantastic precision of the N^{3}LO+N^{4}LL prediction in the highly sensitive low-p_T region. The result is the single most precise experimental determination of $\alpha_s(m_Z)$, and can be seen compared to the world average and other experiments in Fig. 1.

The transverse momentum spectra of the Z and W bosons are also measured in special low pile-up runs at centre-of-mass energies of 5.02 and 13 TeV [4]. The leptonic decays, particularly for the W boson where the neutrino is inferred from transverse momentum imbalance, are much better reconstructed in the cleaner detector environment. The measured distributions can be used to tune simulation predictions and hence reduce modelling uncertainty in future W boson mass measurements.

The $Z \to$ invisible branching ratio (BR) is measured by investigating the ratio of the cross-section for $Z \to$ invisible to that of $Z \to \ell\ell$ in a common phase space [5]. The desired branching ratio is extracted by multiplying the well known $Z \to \ell\ell$ BR the ratio of cross-sections. The result is the single most precise recoil-based measurement of this BR.

The exclusive decays of $W \to \mathcal{M}\gamma$ [6], where $\mathcal{M}$ is a meson, provide a probe to the coupling of the W boson to the different quark generations. Three channels are investigated with the upper limit on $W \to \pi^{\pm}\gamma$ being improved by a factor four compared to the previous best, and the first limits are presented for $W \to K^{\pm}\gamma$ and $W \to \rho^{\pm}\gamma$.

The most abundantly produced electroweak boson at the LHC is the photon, which allows for extremely precise tests of perturbative QCD and a probe of the gluon parton density function (PDF). The most recent ATLAS measurement [7], using the full Run-2 dataset, probes six orders of magnitude in cross-section and measures the cross-section double differentially as a function of the transverse energy and pseudorapidity of the photon. The dependence of the cross-section on the experimentally-imposed photon isolation criteria is also investigated. The measurements are used to test various predictions up to next-to-next-to-leading order in QCD and various PDF sets.

3 Multi-Boson Measurements

The measurements of two- or three-boson final states are of interest due to their connection to bosonic self-couplings that could be portals to new physics. Such effects could be seen as deviations from the predicted cross-sections which are typically parameterised within an Effective Field Theory (EFT) framework.

The most recent measurement of the ZZ process [8] at 13 TeV studies some novel aspects of the underlying physics. Firstly, the polarisation states of the interacting bosons are attempted to be observed in isolation. This is done via a multivariate analysis which attempts to separate the different polarisation states via their kinematic signatures. From the resulting distribution, evidence for the fully longitudinally polarised state is seen with a significance of 4.3σ. Secondly, the analysis exploits an "optimal observable" which is built to be maximally sensitive to CP-odd effects that could arise through anomalous triple gauge couplings. The resulting measurement is used to set limits on the couplings which describe these anomalous interactions.

Measurements of diboson processes with an additional pair of forward, high-mass jets allows for the study of vector boson scattering (VBS) processes, which are a direct probe of quartic gauge couplings. Of the many combinations of two bosons which can be studied, a $W^{\pm}W^{\pm}$ pair has the highest signal to background ratio, but a small cross-section. The most recent ATLAS measurement [9] of this process at

13 TeV measures the cross-section differentially and the resulting measurements are used to set limits on dimension-8 EFT operators and doubly-charged Higgs models. The $W^{\pm}W^{\mp}$ VBS process is also observed for the first time with ATLAS [10]. The measurement uses a neural network to separate the signal from top quark and QCD WW backgrounds, and is observed with a significance of 7.1σ.

Another way to probe quartic couplings is to study the production of triboson systems. Similarly to VBS processes, triboson processes typically have small cross-sections and complicated background sources. The first observations of the $W\gamma\gamma$ [11] and $WZ\gamma$ [12] processes are seen by ATLAS with significances of 5.6σ and 6.3σ respectively. A recent study of $Z\gamma\gamma$ [13] is also presented which performs the first differential cross-section measurements of this process, and uses these to set limits on dimension-8 EFT operators.

4 Run-3 Measurements

The Run-3 of the LHC started in 2022, and continued the high pile-up running at a slightly higher centre-of-mass energy of 13.6 TeV. The first diboson measurement from ATLAS using this dataset has recently been reported [14]. This study of the ZZ process includes fiducial and differential cross-section measurements. The measurement is extrapolated to the total phase space, which allows for comparison to other experiments, as well as at different centre-of-mass energies, and this is shown in Fig. 2. The results confirm the predictability of the SM within this previous unexplored energy regime.

5 Summary

Recent results from the ATLAS experiment are produced using data from the LHC, recorded across many years and in different energy regimes. The measurements are used to stringently test state-of-the-art predictions, however the SM stands up to the test. New, previously unexplored, regions of the SM are also uncovered for the first time. The measurements are used to constrain new physics effects through primarily EFT approaches, which allows us to narrow our search for anomalous effects. This is an important aspect of the physics program of ATLAS that is instrumental in the quest to find new physics.

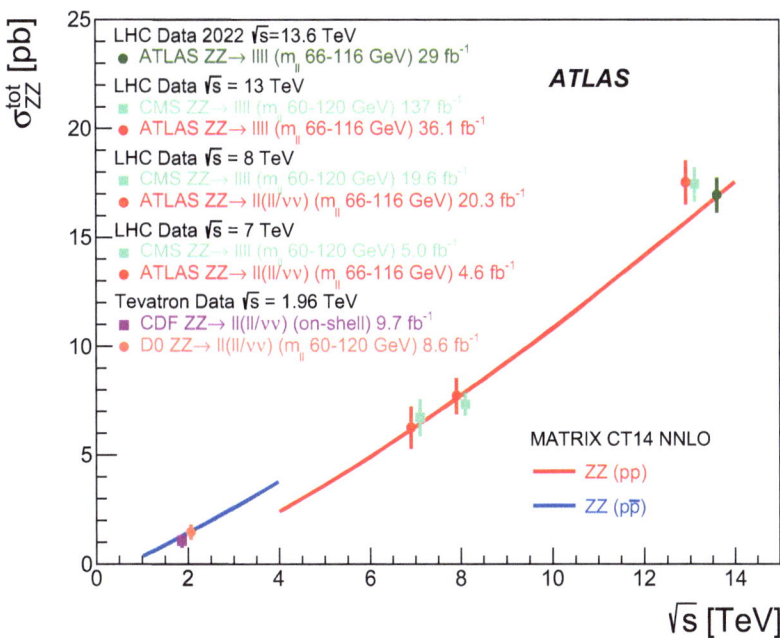

Fig. 2 The measurement of the total cross-section for ZZ production at 13.6 TeV compared to other experiments and at different centre-of-mass energies. The predicted evolution of the cross-section with centre-of-mass energy is also shown. Reprinted under CC-BY-4.0 license from [14]. © 2024, CERN for the benefit of the ATLAS Collaboration

References

1. ATLAS Collaboration, J. Instrum. **3**, S08003 (2008)
2. ATLAS Collaboration, Eur. Phys. J. C **84**, 315 (2024)
3. ATLAS Collaboration, arXiv:2309.12986 [hep-ex]
4. ATLAS Collaboration, Eur. Phys. J. C **84**, 1126 (2024)
5. ATLAS Collaboration, Phys. Lett. B **854**, 138705 (2024)
6. ATLAS Collaboration, Phys. Rev. Lett. **133**, 161804 (2024)
7. ATLAS Collaboration, J. High Energy Phys. **07**, 086 (2023)
8. ATLAS Collaboration, J. High Energy Phys. **12**, 107 (2023)
9. ATLAS Collaboration, J. High Energy Phys. **04**, 026 (2024)
10. ATLAS Collaboration, J. High Energy Phys. **07**, 254 (2024)
11. ATLAS Collaboration, Phys. Lett. B **848**, 138400 (2024)
12. ATLAS Collaboration, Phys. Rev. Lett. **132**, 021802 (2024)
13. ATLAS Collaboration, Eur. Phys. J. C **83**, 539 (2023)
14. ATLAS Collaboration, Phys. Lett. B **855**, 138764 (2024)

Open Access This chapter is licensed under the terms of the Creative Commons Attribution 4.0 International License (http://creativecommons.org/licenses/by/4.0/), which permits use, sharing, adaptation, distribution and reproduction in any medium or format, as long as you give appropriate credit to the original author(s) and the source, provide a link to the Creative Commons license and indicate if changes were made.

The images or other third party material in this chapter are included in the chapter's Creative Commons license, unless indicated otherwise in a credit line to the material. If material is not included in the chapter's Creative Commons license and your intended use is not permitted by statutory regulation or exceeds the permitted use, you will need to obtain permission directly from the copyright holder.

Search for Dark Matter with 2HDM+a in pp Collisions at $\sqrt{s}$ = 13 TeV with the ATLAS Detector

Sanae Ezzarqtouni

1 Introduction

Despite its experimental successes, the Standard Model (SM) falls short in explaining dark matter, a significant component of the universe's energy. An alternative beyond standard model suggests a stable, neutral, weakly-interacting massive particle (WIMP or χ) as a viable dark matter candidate. Various methods, including indirect searches, direct detection, and collider experiments are used to study dark matter and its properties.

WIMP production at particle colliders is identified by observing significant missing transverse energy (E_T^{miss}) in association with SM particles with "E_T^{miss}+X" signatures, where the X represents objects that interact with the detector. An additional strategy involves searching for visible decays of mediator particles to constrain dark matter models.

These searches are analyzed within the context of an extra pseudo-scalar mediator, incorporated into the Two Higgs Doublet Model (2HDM), identified as 2HDM+a [1]. This mediator facilitates interactions between the visible and dark sectors. The model is defined by 14 parameters, yet only 5 parameters, which include the masses of m_A, m_a, m_χ, $\tan \beta$ and $\sin \theta$, are treated as free parameters. The choise of free parameters and the associated benchmark parameter scans were comprehensively examined and documented within the white paper [1].

Sanae Ezzarqtouni on behalf of the ATLAS Collaboration

S. Ezzarqtouni (✉)
Hassan II University of Casablanca, Faculty of Sciences Ain Chock, Casablanca, Morocco
e-mail: sanae.ezzarqtouni@cern.ch

2 Input Analyses

A variety of searches, examining different final states and addressing both invisible and visible mediator decays, explore the 2HDM+a model. None of these searches revealed a deviation from the SM prediction. Consequently, their statistical combination is used to put constraints on 2HDM+a (Fig. 1).

2.1 Searches for Invisible Final States: $X + E_T^{miss}$ Searches

Z(ll)+E_T^{miss} [2]: The signal is characterized by a pair of leptons originating from Z-boson decay, accompanied by substantial missing transverse energy in the recoil. The event selection criteria are based on the observation of significant E_T^{miss} and a pair of leptons with high transverse momentum. Specifically, two opposit-sign leptons of same flavor with a $p_T > 30\ (20)$ GeV for the leading (subleading) lepton. Three Control Regions (CRs) are employed to constrain the SM background predictions: the 3-lepton CR, 4-lepton CR, and eμ CR.

h(bb)+E_T^{miss} [3]: The analysis signature involves two b-jets, arising from the decays of a SM Higgs boson and significant E_T^{miss}, denoted as E_T^{miss}+h(bb). The analysis utilizes two b-jets reconstruction techniques, namely merged and resolved selections. The resolved analysis is applied for $E_T^{miss} < 500$ GeV, while the merged

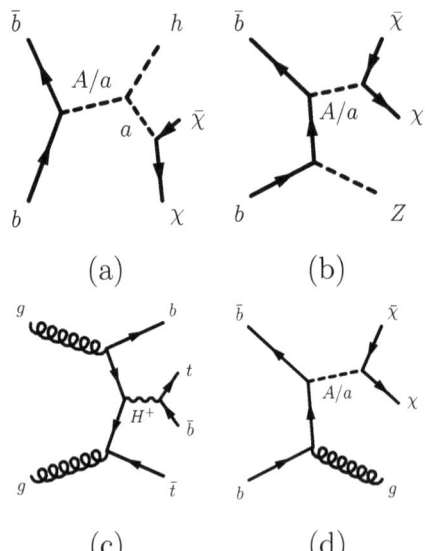

Fig. 1 Representative Feynman diagrams for (**a**): $E_T^{miss} + h$ signature, the $E_T^{miss} + Z$ signature (**b**), $tbH^{\pm}(tb)$ (**c**) and the $E_T^{miss} + j$ signature (**d**) in the 2HDMa

analysis is used for $E_T^{miss} > 500$ GeV. The main backgrounds stem from $t\bar{t}$ and the production of a weak boson in association with jets, with additional contributions from single-top, diboson, and SM Vh production.

2.2 Searches for Visible Final States

tb$H^\pm$(tb) [4]: This search targets the production of heavy charged Higgs bosons, $H^\pm$ decaying into top and bottom quarks, with masses ranging from 200 to 2000 GeV. Events are subjected to preselection criteria involving single-lepton triggers, which require exactly one electron or one muon with $p_T > 27$ GeV and at least five jets with $p_T > 25$ GeV. The analysis mandates the identification of at least three of the jets as b-jets. among the reconstructed jets, The events are classified into four distinct regions, labeled as "5j,3b", "5j, $\geq$ 4b", " $\geq$ 6j, 3b", and " $\geq$ 6j, $\geq$ 4b". The primary backgrounds in this search predominantly consist of events from $t\bar{t}$+jets and single top-quark production in the Wt channel.

3 Summary of the Constraints on the 2HDM+a

The most sensitive analyses from the 2HDM+a model, including Z(ll)+E_T^{miss}, h(bb)+E_T^{miss}, and tb$H^\pm$(tb) are statistically combined. The resulting combined limit consistently exhibits greater stringency than the limits obtained from individual analyses. The statistical combination is facilitated by the independence of the three analyses. The Z($\to$ ll)+E_T^{miss} analysis includes a b-jet veto, while the h($\to$bb)+E_T^{miss} and tb$H^\pm$(tb) analyses require a minimum of two and three b-jets, respectively. Moreover, the selections of the tb$H^\pm$(tb) signal region require a charged lepton, which is vetoed by the h($\to$bb)+E_T^{miss} selections. Consequently, no overlap is expected between the signal selections of the three analyses. Furthermore, the combination of analyses is achieved by constructing their combined likelihood and optimizing the corresponding profile likelihood ratio [5].

The exclusion contours for the $m_A - m_a$ and $tan\beta - m_a$ scans are depicted in Fig. 2. The scans consider $\sin\theta$ values of 0.35 and 0.7. Additionally, the exclusion contours are presented as a function of m_χ and $\sin\theta$. The regions within the observed contours are excluded by each analysis and the Combination. The figures indicate that the sensitivities are further improved by the combination.

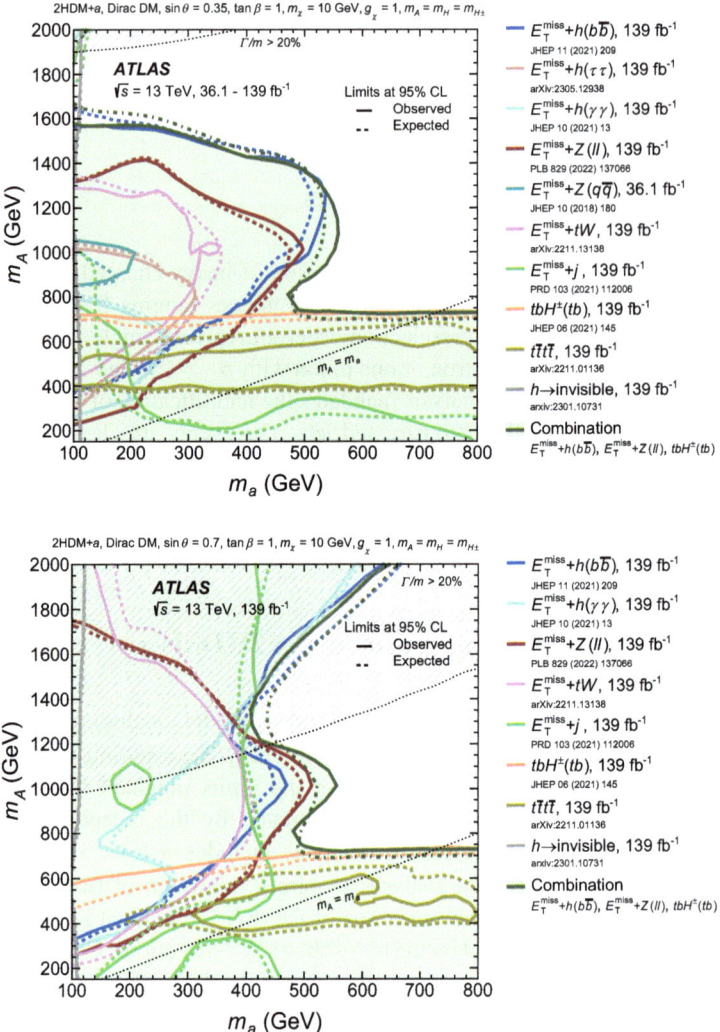

Fig. 2 The exclusion regions at a 95% confidence level (CL) are shown in solid lines for the observed results and dashed lines for the expected results. These exclusion regions are presented in two planes: (top) the (m_a, m_A) plane and (middle) the $(m_a, \tan\beta)$ plane. Additionally, the exclusion regions are shown as a function of m_χ (left) and as a function of $\sin\theta$ (right) in the bottom panel. These results demonstrate the statistical combination of the $Z(ll)+E_T^{miss}$, $h(bb)+E_T^{miss}$, and $tbH^\pm(tb)$ searches. Reprinted under CC-BY-4.0 from [5]. © The Author(s)

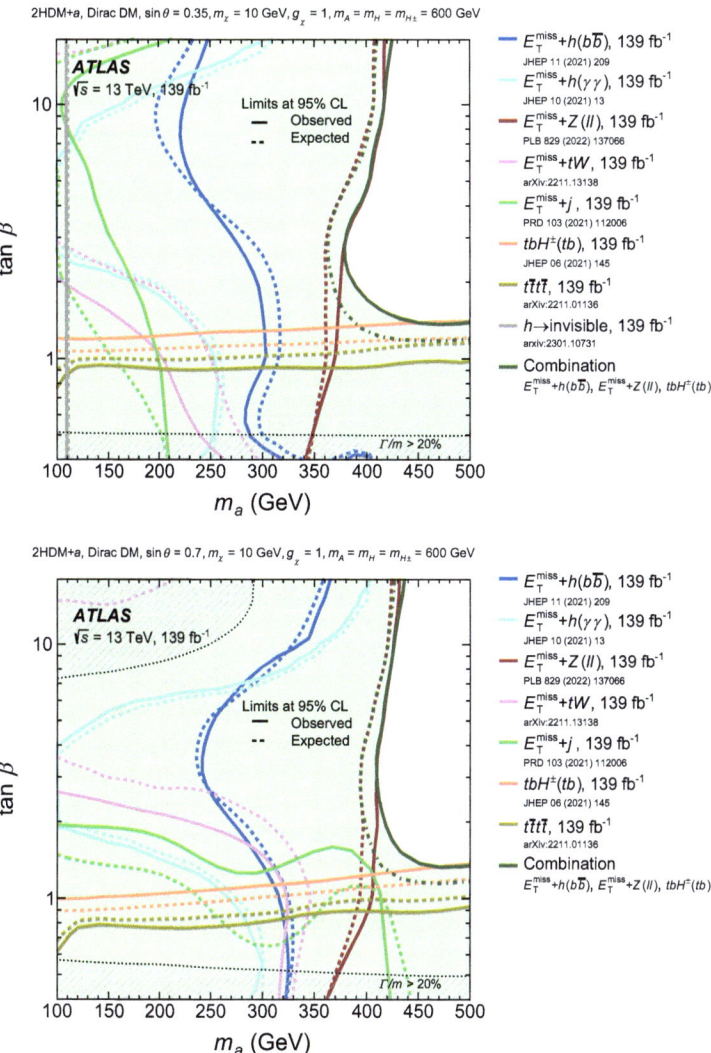

Fig. 2 (continued)

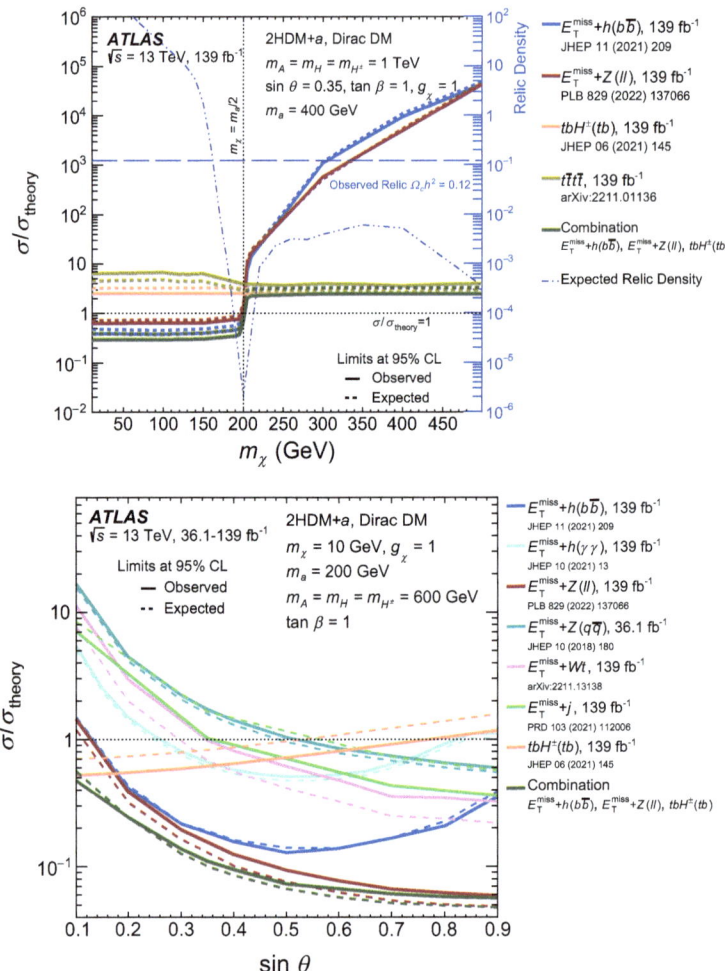

Fig. 2 (continued)

References

1. T. Abe et al., LHC dark matter working group: next-generation spin-0 dark matter models. J. Phys. Dark Univ. **27**, 100351 (2019). https://doi.org/10.1016/j.dark.2019.100351
2. ATLAS Collaboration, Search for associated production of a Z boson with an invisibly decaying Higgs boson or dark matter candidates at $\sqrt{s} = 13$ TeV with the ATLAS detector. Phys. Lett. B **829**, 137066 (2021). https://doi.org/10.1016/j.physletb.2022.137066
3. ATLAS Collaboration, Search for dark matter produced in association with a Standard Model Higgs boson decaying into b-quarks using the full Run 2 dataset from the ATLAS detector. J. High Energy Phys. **11**, 209 (2021). https://doi.org/10.1007/JHEP11(2021)209

4. ATLAS Collaboration, Search for charged Higgs bosons decaying into a top quark and a bottom quark at $\sqrt{s} = 13$ TeV with the ATLAS detector. J. High Energy Phys. **06**, 145 (2021). https://doi.org/10.1007/JHEP06(2021)145
5. ATLAS Collaboration, Combination and summary of ATLAS dark matter searches interpreted in a 2HDM with a pseudo-scalar mediator using 139 fb^{-1} of $\sqrt{s} = 13$ TeV pp collision data (2023). https://doi.org/10.48550/arXiv.2306.00641

Open Access This chapter is licensed under the terms of the Creative Commons Attribution 4.0 International License (http://creativecommons.org/licenses/by/4.0/), which permits use, sharing, adaptation, distribution and reproduction in any medium or format, as long as you give appropriate credit to the original author(s) and the source, provide a link to the Creative Commons license and indicate if changes were made.

The images or other third party material in this chapter are included in the chapter's Creative Commons license, unless indicated otherwise in a credit line to the material. If material is not included in the chapter's Creative Commons license and your intended use is not permitted by statutory regulation or exceeds the permitted use, you will need to obtain permission directly from the copyright holder.

Jets Physics Simulations and Analysis with the ATLAS Experiment

Saad El Farkh

1 Introduction

Multijet measurements are integral to advancing our understanding of Quantum Chromodynamics (QCD). By analyzing the complex interplay of multiple jets in proton-proton collisions, these studies offer critical insights into both theoretical and experimental aspects of particle physics. Specifically, multijet events serve as robust tests for perturbative QCD (pQCD) calculations across different levels of precision—Leading Order (LO), Next-to-Leading Order (NLO), and Next-to-Next-to-Leading Order (NNLO)—while also shedding light on parton shower modeling and aiding in the determination of fundamental QCD parameters.

This paper provides an overview of recent multijet studies conducted by the ATLAS Collaboration [1] at the Large Hadron Collider (LHC), focusing on three key analyses at a center-of-mass energy of 13 TeV:

1. The study of event shapes in high-p_T multijet events [2], which characterizes the jet distributions and their geometrical properties.
2. The determination of the strong coupling constant (α_s) [3] using transverse energy-energy correlations (TEEC) in multijet events, offering a precision measurement of this fundamental QCD parameter.

Copyright 2024 CERN for the benefit of the ATLAS Collaboration. CC-BY-4.0 license.
Saad El Farkh on behalf of the ATLAS Collaboration.

S. El Farkh (✉)
Faculty of Sciences, University Ibn Tofail, Kenitra, Morocco
e-mail: Saad.el.farkh@cern.ch

© The Author(s) 2026
Y. Tayalati, M. Gouighri (eds.), *The First African Conference on High Energy Physics*, Springer Proceedings in Physics 425,
https://doi.org/10.1007/978-3-031-88933-2_13

3. The investigation of multijet event isotropies [4], which introduces novel observables to probe the isotropic nature of QCD radiation, compared to advanced Monte Carlo simulations.

2 Event Shapes in High-p_T Multijet Events at 13 TeV

Event shape observables provide insight into the topology and energy distribution of multijet events. In this analysis, we focus on transverse thrust ($\tau_\perp$), which describes the geometry of jet events in the transverse plane. Figure 1 provides a visualization of event shapes for different jet multiplicities.

We also measure transverse sphericity ($S_\perp$) across different jet multiplicities and transverse momentum scales (H_{T2}), comparing data and MC predictions. The comparison reveals discrepancies between data and simulations, particularly at low jet multiplicities, but an improvement in agreement as the jet multiplicity increases (Fig. 2).

These event shape measurements are crucial for testing and refining QCD predictions and simulations in high-p_T multijet events, revealing discrepancies between data and Monte Carlo predictions that vary with jet multiplicity and simulation model.

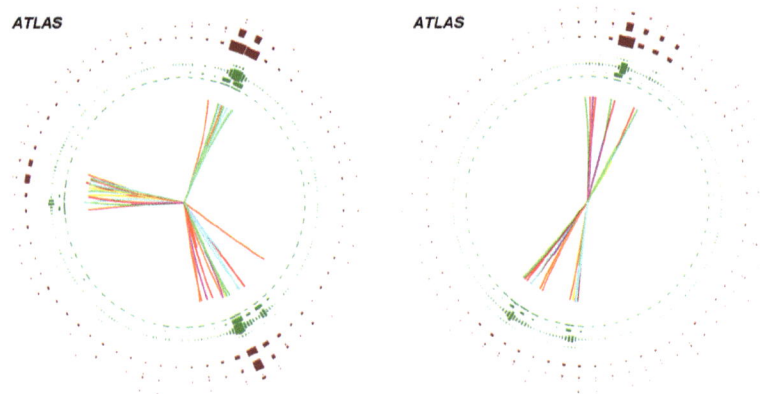

Fig. 1 Transverse plane projection of a three-jet event with high values of $\tau_\perp$ and $S_\perp$ (left), and a five-jet event with low values $\tau_\perp$ and $S_\perp$ (right). The colours are chosen for illustrative purposes. **Reprinted under CC-BY-4.0 license from [2]. © 2021, CERN for the benefit of the ATLAS Collaboration**

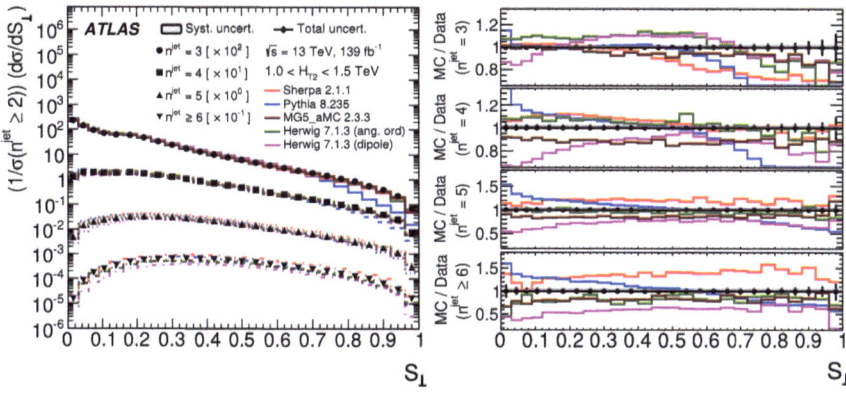

Fig. 2 Comparison between data and MC predictions as a function of $S_\perp$ for different jet multiplicities. The right panel shows the ratio between MC and data, with uncertainty bands indicating the systematic uncertainty. Reprinted under CC-BY-4.0 license from [2]. © 2021, CERN for the benefit of the ATLAS Collaboration

3 Determination of the Strong Coupling Constant in Multijet Events

Multijet events play a crucial role in measuring QCD event shapes and determining the strong coupling constant (α_s), which influences jet emission rates and event distributions. Key observables include the Energy-Energy Correlation (EEC), reflecting angular energy distributions, the Transverse Energy-Energy Correlation (TEEC), focusing on the transverse plane, and the Asymmetry of TEEC (ATEEC), comparing forward and backward correlations.

This analysis uses ATLAS Run-2 data (139 fb^{-1}, $\sqrt{s}$ = 13 TeV) with strict cuts: $p_T > 60$ GeV, $|\eta| < 2.4$, and $H_{T2} > 1$ TeV. Results compare TEEC and ATEEC with Monte Carlo predictions, providing insights into α_s-sensitive models and improving parton shower accuracy (Fig. 3).

The strong coupling constant $\alpha_s(Q)$ is extracted from the TEEC and ATEEC measurements by fitting the data across ten H_{T2} intervals, which represent different energy scales. These fits provide precise values for $\alpha_s(m_Z)$, as summarized below.

$$\alpha_s(m_Z) = 0.1175 \pm 0.0006 \text{ (exp.)}^{+0.0034}_{-0.0017} \text{ (theo.) and}$$

$$\alpha_s(m_Z) = 0.1185 \pm 0.0009 \text{ (exp.)}^{+0.0025}_{-0.0012} \text{ (theo.)}$$

These values are consistent with the world average and previous measurements from ATLAS and other experiments. NNLO corrections to three-jet production significantly reduce the theoretical uncertainties. The measured $\alpha_s(Q)$ values are compared with the QCD predictions using the RGE, showing excellent agreement. The measurements of $\alpha_s(m_Z)$ from TEEC and ATEEC agree with the world average

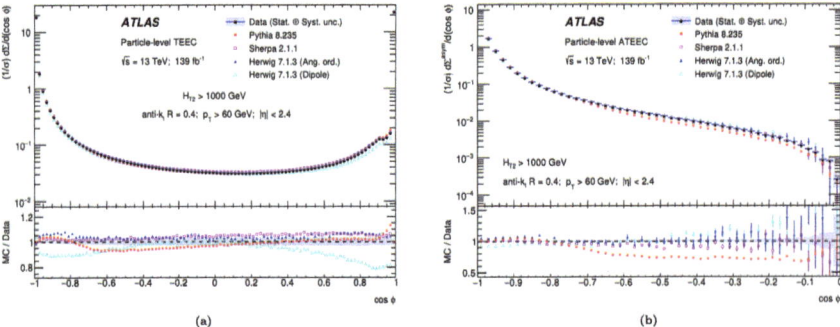

Fig. 3 Comparison of data and MC predictions for (**a**) TEEC and (**b**) ATEEC in the inclusive H_{T2} bin. Data error bands include statistical and systematic uncertainties. Lower panels show MC/data ratios, with statistical uncertainties. Negative ATEEC values are excluded in logarithmic scale. Reprinted under CC-BY-4.0 license from [3]. © 2023, CERN for the benefit of the ATLAS Collaboration

and theoretical predictions, providing a precise determination of α_s with reduced uncertainties.

4 Multijet Event Isotropies at 13 TeV

In this study, multijet event isotropies are analyzed using the Energy-Mover's Distance (EMD), which measures the "work" required to transport one event configuration to another with equal energy. The focus is on multijet events with $N_{jet} > 2$ and $H_{T2} > 400$ GeV, using data from the ATLAS detector at $\sqrt{s} =$ 13 TeV. The EMD is applied to three reference geometries:

1. **Thrust:** EMD from an event to a dipole configuration.
2. **2D Event Isotropy:** EMD from an event to a ring.
3. **3D Event Isotropy:** EMD from an event to a cylinder.

These geometries characterize the spatial distribution of energy and momentum in multijet events, enhancing our understanding of QCD dynamics and BSM physics. The ring-like isotropy is dominated by dijet and nearly isotropic multijet events, with discrepancies in MC predictions, especially with Powheg. The cylindrical isotropy is influenced by dijets in the forward region and multijets across the rapidity-azimuth plane, with MC predictions failing to match data, and parton shower models playing a larger role than hadronization models (Fig. 4).

These measurements enhance our understanding of multijet event shapes and provide valuable constraints for improving Monte Carlo simulations. The differences between parton shower models and the sensitivity of isotropy measurements to the number of jets highlight the need for further refinement of these models to better describe high-energy hadronic collisions at the LHC.

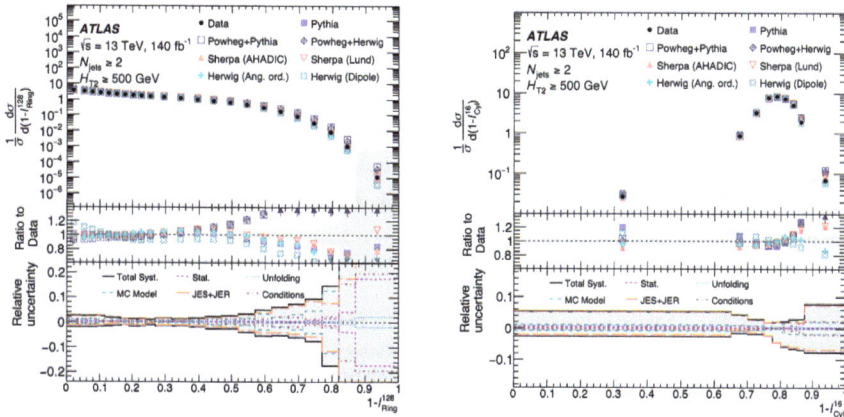

Fig. 4 Comparison of Ring-like and Cylindrical Event Isotropy Distributions in Data and MC Predictions. **Reprinted under CC-BY-4.0 license from [4]. © 2023, CERN for the benefit of the ATLAS Collaboration**

These isotropy measurements offer a unique opportunity to test QCD predictions, refine event generators, and probe potential BSM effects in multijet final states, ultimately improving the precision of simulations and the sensitivity of future searches for new physics.

5 Conclusion

Recent ATLAS studies on multijet events at 13 TeV have provided significant insights into the dynamics of QCD and have the potential to improve the precision of event generators. By introducing novel observables like event isotropies and transverse energy-energy correlations (TEEC), these studies offer new tools to better describe the complexities of high-energy hadronic collisions. The determination of the strong coupling constant (α_s) from multijet events has been achieved with high accuracy, contributing to fundamental QCD measurements. Additionally, the application of optimal transport techniques to analyze event isotropies marks a new milestone in collider physics. With new results expected from Run 3, these studies will continue to refine our understanding of multijet final states and probe new physics.

References

1. The ATLAS Collaboration, The ATLAS experiment at the CERN Large Hadron Collider. J. Instrum. **3**, S08003 (2008). https://doi.org/10.1088/1748-0221/3/08/S08003

2. The ATLAS Collaboration, Measurement of hadronic event shapes in high-p_T multijet final states at $\sqrt{s} = 13$ TeV with the ATLAS detector. J. High Energy Phys. **01**, 188 (2021). https://doi.org/10.1007/JHEP01(2021)188
3. The ATLAS Collaboration, Measurement of hadronic event shapes in high-p_T multijet final states at $\sqrt{s} = 13$ TeV with the ATLAS detector. J. High Energy Phys. **07**, 085 (2023). https://doi.org/10.1007/JHEP07(2023)085
4. The ATLAS Collaboration, Measurements of multijet event isotropies using optimal transport with the ATLAS detector. J. High Energy Phys. **10**, 060 (2023). https://doi.org/10.1007/JHEP10(2023)060

Open Access This chapter is licensed under the terms of the Creative Commons Attribution 4.0 International License (http://creativecommons.org/licenses/by/4.0/), which permits use, sharing, adaptation, distribution and reproduction in any medium or format, as long as you give appropriate credit to the original author(s) and the source, provide a link to the Creative Commons license and indicate if changes were made.

The images or other third party material in this chapter are included in the chapter's Creative Commons license, unless indicated otherwise in a credit line to the material. If material is not included in the chapter's Creative Commons license and your intended use is not permitted by statutory regulation or exceeds the permitted use, you will need to obtain permission directly from the copyright holder.

Reconstructing Long-Lived Particles with the ILD Detector

Jan Klamka

1 Introduction

The concept of Beyond the Standard Model (BSM) particles with macroscopic lifetimes has been extensively studied in recent years, also in the context of collider searches, as the LLPs can naturally appear in many BSM scenarios [1].

The ILD [2] is a multipurpose detector concept for an experiment at the Higgs factory, originally proposed for the International Linear Collider (ILC) [3]. The ILD tracking systems feature vertex detector (three double silicon layers) and a silicon inner tracker (two double silicon layers), extended by seven silicon discs in the forward region. They are surrounded by a time projection chamber (TPC), the main tracker with an inner radius of 329 mm and an outer radius of 1770 mm.

2 Analysis Framework

Two extreme scenarios were selected as benchmarks, based on their kinematic properties, and not as preferred points in a model parameter space. The more challenging case involves heavy scalars production with small-boost and a non-

---

Supported by the National Science Centre (Poland) under OPUS research project no. 2021/43/B/ST2/01778.
Jan Klamka on behalf of the ILD Concept Group.

J. Klamka (✉)
Faculty of Physics, University of Warsaw, Warsaw, Poland
e-mail: jan.klamka@fuw.edu.pl

pointing, low-p_T track pair in the final state, as exemplified by the Inert Doublet Model (IDM). IDM introduces four additional scalars, including two neutral ones, A and H (where the latter is a stable dark matter candidate). They can be produced in $e^+e^- \to$ AH process, with A $\to Z^{(*)}$H as the main decay channel. If the mass splitting, $\Delta m_{AH} = m_A - m_H$, is sufficiently small, A could be long-lived [4]. Four IDM benchmarks with $\Delta m_{AH} = 1, 2, 3, 5$ GeV, $c\tau$ of A fixed to 1 m, and $m_A = 155$ GeV are analysed. Only $Z^* \to \mu\mu$ decays are considered.

The second scenario involves the production of a light pseudoscalar, specifically an axion-like particle (ALP). ALPs, with masses $\mathcal{O}(1\,\text{GeV})$ and macroscopic lifetimes, could be produced at e^+e^- colliders in $e^+e^- \to a\gamma$ channel [5]. Four benchmarks with different ALP masses ($m_a = 0.3, 1, 3, 10$ GeV) and decay lengths $c\tau = 10 \cdot m_a$ mm/GeV are analyzed, to maintain consistency in the number of decays within the detector volume. Only $a \to \mu\mu$ decays are considered.

WHIZARD was used to generate the IDM (ALP) samples at 500 GeV (250 GeV) ILC, with GEANT4 for full detector simulation. For vertex finding, an algorithm that uses a calculated distance between the track helices was employed, where the vertex is placed in between the points of the closest approach of the helices.

3 Background and Its Reduction

At linear colliders, beam-induced background events contribute in each bunch-crossing (BX). At the ILC, on average 1.05 events of $\gamma\gamma \to$ hadrons and a fraction of incoherent e^+e^- pairs occur per single BX. This background can overlay on hard events affecting reconstruction, but for low-p_T final states in the signal channel, it can be standalone background and must be considered separately.

Overlay events may contain numerous low-p_T tracks starting near the IP. The algorithm designed for this study, finds many (mostly fake) vertices in this busy environment, especially near the beam axis. Tracks in overlay events can also resemble kinematically those in the considered signal scenarios. Given the order of 10^{11} BXs expected at the ILC, overlay becomes a significant background. It is still possible to strongly suppress overlay background maintaining the presented model-independent approach. Selection criteria, including veto against decays of K^0, Λ^0, and γ conversions, track-pair $p_T > 1.9$ GeV, and cuts on track first hit and "helix-circle" centers distances, along with preselection to suppress fake vertices, achieve rejection factor of 10^9 (10^{10}) for $\gamma\gamma \to$ hadrons (e^+e^- pairs).

4 Results in the TPC

The vertex finding efficiency, defined as the ratio of the number of correctly reconstructed vertices to the number of decays inside TPC, varies within 37–60% range in the heavy scalars case, for scenarios with $\Delta m_{AH} > 1$ GeV. The

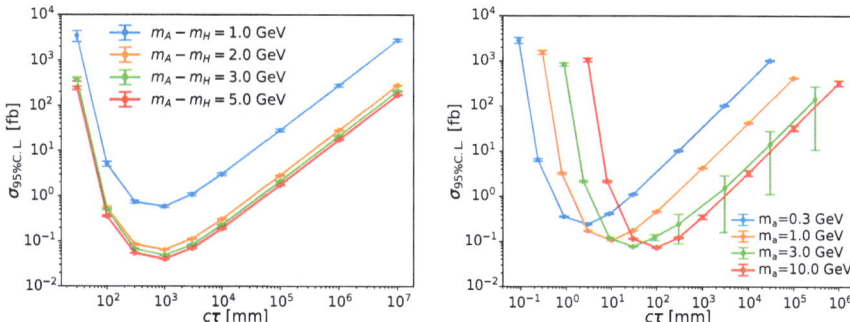

Fig. 1 Expected 95% C.L. upper limits on the signal production cross-section for considered benchmarks and different LLP mean decay lengths for the heavy scalars at $\sqrt{s} = 500\,\text{GeV}$ (left) and the light pseudoscalar at $\sqrt{s} = 250\,\text{GeV}$ (right)

reconstructed vertex was considered to be "correct", if its distance from the true vertex was less than 30 mm. For the $m_{AH} = 1\,\text{GeV}$ scenario the efficiency was only 4% and it could be improved using a dedicated approach [6]. The purity for all benchmarks was higher than 99%. The vertex finding efficiency for heavy scalars is higher for decays in TPC than for decays in the inner tracker region.

The corresponding efficiency for the light pseudoscalar case was 54–78% for scenarios with $m_a \geq 1\,\text{GeV}$, reaching purity of 83–99%. For the $m_a = 300\,\text{MeV}$ benchmark, efficiency drops to 24% with 43% purity. Contrary to the heavy scalar case, the efficiency is higher for decays inside the inner silicon tracker, and it increases with the LLP mass (decreasing boost). For higher boosts two tracks in the final state become almost colinear and the separation of hits close to the vertex is better in the silicon detector, providing much higher resolution.

An estimate was also made of the expected 95% C.L. limit on the signal production cross-section. An integrated luminosity of $2\,\text{ab}^{-1}$ was assumed at 250 GeV ILC, corresponding to 10 years of data collection, and $4\,\text{ab}^{-1}$ collected at 500 GeV in 8.5 years. With 10^{11} BXs each year, 1.05 (1.00) $\gamma\gamma \to$ hadrons (e^+e^- pair) events expected per BX, and with the total background rejection of 10^{-9} (10^{-10}), one can expect around 1150 background events. Figure 1 presents the limits for different benchmark scenarios and a range of the LLP mean decay lengths $c\tau$. For short lifetimes, a small number of LLP decays within the TPC volume results in worse sensitivities, comparable to scenarios with very high $c\tau$.

5 All-Silicon ILD Design

An alternative ILD design was also considered, with the TPC replaced by the outer silicon tracker modified from CLICdet [7] detector model. As for CLICdet, the Conformal Tracking [8] algorithm was also used as a pattern recognition tool.

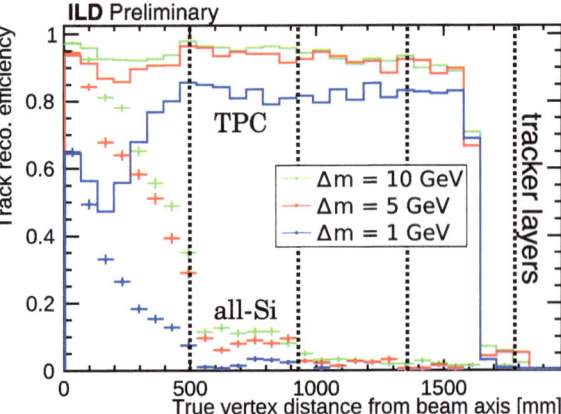

Fig. 2 Track reconstruction efficiency as a function of distance from the beam axis for the heavy scalars in the all-silicon ILD design (points) and the baseline (solid lines). Different colours correspond to the scalar mass-splitting scenarios. Vertical dashed lines show the position of outer barrel layers in all-silicon model. Reprinted under CC-BY-4.0 license from [9]. © 2023, The Author(s)

The tracking efficiency is shown in Fig. 2 for heavy scalars, as a function of the decay vertex distance from the beam axis, and compared with the standard ILD design. If LLP decays inside the vertex detector or inner tracker (same in both designs), the performance is similar. However, for the all-silicon design, the efficiency drops rapidly for larger distances, while the standard ILD design maintains high efficiency throughout the tracker volume. Limited number of layers and hits required for track reconstruction contribute to the rapid efficiency decline for the silicon tracker, reaching almost zero at 1 m from the beam axis.

6 Conclusions

The study explores prospects of LLP searches in the ILD using displaced vertex signature and a model-independent approach. Two challenging signatures and background from beam interactions are considered. The estimated limit on the signal production cross-section below 1 fb is reached for several benchmarks within the $c\tau$ range of 1–10^5 mm. Tests using an all-silicon ILD design emphasize the TPC's crucial role in sensitivity to LLP searches with soft final states.

References

1. L. Lee et al., Collider searches for long-lived particles beyond the standard model. Prog. Part. Nucl. Phys. **106**, 210–255 (2019). https://doi.org/10.1016/j.ppnp.2019.02.006 [Erratum: Prog. Part. Nucl. Phys. 122, 103912 (2022)]
2. H. Abramowicz et al., International large detector: interim design report (2020). arXiv:2003.01116
3. P. Bambade et al., The international linear collider: a global project (2019). arXiv:1903.01629

4. J. Kalinowski et al., IDM benchmarks for the LHC and future colliders. Symmetry **13**(6), 991 (2021). https://doi.org/10.3390/sym13060991
5. R. Schäfer et al., Near or far detectors? A case study for long-lived particle searches at electron-positron colliders. Phys. Rev. D **107**(7), 076022 (2023). https://doi.org/10.1103/PhysRevD.107.076022
6. M. Berggren et al., Tackling light higgsinos at the ILC. Eur. Phys. J. C **73**(12), 2660 (2013). https://doi.org/10.1140/epjc/s10052-013-2660-y
7. N. Alipour Tehrani et al., CLICdet: The post-CDR CLIC detector model (2017). http://cds.cern.ch/record/2254048
8. E. Brondolin et al., Conformal tracking for all-silicon trackers at future electron–positron colliders. Nucl. Instrum. Meth. A **956**, 163304 (2020). https://doi.org/10.1016/j.nima.2019.163304
9. J. Klamka, Reconstructing long-lived particles with the ILD detector. PoS **EPS-HEP2023**, 458 (2024). https://doi.org/10.22323/1.449.0458

Open Access This chapter is licensed under the terms of the Creative Commons Attribution 4.0 International License (http://creativecommons.org/licenses/by/4.0/), which permits use, sharing, adaptation, distribution and reproduction in any medium or format, as long as you give appropriate credit to the original author(s) and the source, provide a link to the Creative Commons license and indicate if changes were made.

The images or other third party material in this chapter are included in the chapter's Creative Commons license, unless indicated otherwise in a credit line to the material. If material is not included in the chapter's Creative Commons license and your intended use is not permitted by statutory regulation or exceeds the permitted use, you will need to obtain permission directly from the copyright holder.

The Cross Section of Inverse Beta Decay

Giulia Ricciardi ⓘ **, N. Vignaroli** ⓘ **, and F. Vissani** ⓘ

1 Introduction

The interaction process

$$\bar{\nu} + p \to e^+ + n \tag{1}$$

is known as inverse β decay (IBD). Its cross section was estimated for the first time by Bethe and Peierls in 1934 [1], using the Fermi theory, and gave an extremely small value ($\sigma \approx 10^{-44}$ cm^2 at $E(\bar{\nu}) = 2$ MeV), leading them to believe that the neutrino is an "undetectable particle". It is kind of ironic that in 1956 the inverse beta decay was exactly the process that allowed the first direct experimental evidence of

---

G. Ricciardi (✉)
Dipartimento di Fisica E. Pancini, Università di Napoli Federico II, Naples, Italy

INFN, Naples, Italy

MSA, Naples, Italy
e-mail: giulia.ricciardi2@unina.it; giulia.ricciardi@na.infn.it

N. Vignaroli
Dipartimento di Matematica e Fisica E. De Giorgi, Università del Salento, Lecce, Italy

INFN, Lecce, Italy
e-mail: natascia.vignaroli@unisalento.it

F. Vissani
INFN Laboratori Nazionali del Gran Sasso, L'Aquila, Italy
e-mail: francesco.vissani@lngs.infn.it

© The Author(s) 2026
Y. Tayalati, M. Gouighri (eds.), *The First African Conference on High Energy Physics*, Springer Proceedings in Physics 425,
https://doi.org/10.1007/978-3-031-88933-2_15

neutrinos [2].[1] Since then, the IBD has always played a central role in neutrino physics, giving rise to a positron and a neutron, both potentially observable, and being in common use at energies below ~ 20 MeV and its wide use in water- or hydrocarbon-based detectors, which are relatively cheap materials and rich in target protons. As a consequence, an accurate estimate of its cross section is essential for the accuracy of the results in several and different neutrino experiments. In this work, we discuss its accurate evaluation [4] and relevance in the current experimental framework.

2 The Cross Section Evaluation

We summarize the main steps needed to build the IBD cross section:[2]

1. write down the matrix element of the charged weak current between proton and neutron states, which in the most general form depends on six dimensionless Lorentz invariant form factors:

$$
\mathcal{J}_\mu = \bar{u}_n \left(f_1 \gamma_\mu + g_1 \gamma_\mu \gamma_5 + i f_2 \sigma_{\mu\nu} \frac{q^\nu}{2M} + g_2 \frac{q_\mu}{M} \gamma_5 \right.
$$
$$
\left. + f_3 \frac{q_\mu}{M} + i g_3 \sigma_{\mu\nu} \frac{q^\nu}{2M} \gamma_5 \right) u_p; \tag{2}
$$

2. calculate the tree level amplitude for the IBD, by using the lepton current and the W propagator. By squaring and averaging, one obtains the differential cross section

$$
\frac{d\sigma}{dt} = \frac{G_F^2 \cos^2 \theta_C}{64\pi (s - m_p^2)^2} |\mathcal{M}^2| \tag{3}
$$

where G_F is the Fermi coupling, the parameters $s = (p_\nu + p_p)^2$, $t = (p_\nu - p_e)^2$ are Mandelstam variables and θ_C is the Cabibbo angle (that is linked to the $u - d$ element of the CKM matrix by the equality $\cos \theta_C = V_{ud}$);

3. include the radiative corrections at leading order [5] using:

$$
d\sigma(E_\nu, E_e) \to d\sigma(E_\nu, E_e) \left[1 + \frac{\alpha}{\pi} \left(6.00 + \frac{3}{2} \log \frac{m_p}{2E_e} + 1.2 \left(\frac{m_e}{E_e} \right)^{1.5} \right) \right] \tag{4}
$$

---

[1] For a brief historical survey see e.g. [3].
[2] For details see Ref. [4].

where α is the fine-structure constant. Next order corrections and other effects such as isospin breaking are estimated to be small and neglected in the low range of neutrino energies (below $\sim$10 MeV).

A new series of modern calculations was started by Vogel and Beacom in 1999 [6] and continued by Strumia and Vissani [7] in 2002. An updated and accurate assessment of the uncertainties was addressed in 2023 by Ricciardi, Vignaroli and Vissani in 2023 [4].

The estimation of theoretical uncertainties is crucial in experimental measurements aiming at high precision. In the cross section built as outlined before, the uncertainties are reflection of the uncertainties in the knowledge of the form factors and the Cabibbo angle.

The hadronic current $\mathcal{J}_\mu$ includes the vector and axial vector terms with form factors f_3 and g_3, that transform as second class currents under G-parity, according to a classification due to Weinberg [8]. G-parity is a symmetry for strong interactions, broken by mass differences. It turns out that second class currents can be safely neglected at the current level of precision [4].

The input parameters which determine the leading uncertainties in the cross section evaluation vary with the energy range; they are

- the Cabibbo angle and the axial coupling $g_1(0)$ (lowest energies);
- the axial radius r_A (higher energies).

The magnitude of the mixing element $V_{ud} = \cos\theta_C$ can be derived directly from the super-allowed $0^+ \to 0^+$ nuclear beta decays, which are pure vector transitions. The result is in tension with the one derived indirectly from the CKM unitarity, using the currently accepted values for $|V_{us}|$ and $|V_{ub}|$. That has been interpreted as a failure of unitarity of around 2σ and possibly more [9]. We believe it is reasonable to assume that this difference merely signals limits to the available interpretations and measurements, and we combine the data, enlarging all errors contributing to the result by the scale factor $S = \sqrt{\chi^2/(N-1)} = 2.0$ for a conservative estimation of the uncertainty, in agreement with the PDG suggestion [10]. We obtain $V_{ud} = 0.9743(3)$.

There are eight different measurements of the normalized axial coupling $\lambda = -g_1(0)/f_1(0)$ using decay polarized neutrons, with the latest measurement, published in 2019 by the Perkeo III collaboration [11], being much more precise than the others. In a conservative perspective, we include all results, but enlarging their error by a factor 2 to take into account discrepancies among older and newer results. That yields the value $\lambda = 1.2760(5)$, which is within 1σ from Ref. [11] and agrees with the global average.

The theoretical relation [12]

$$\frac{1}{\tau_n} = \frac{V_{ud}^2(1+3\lambda^2)}{4906.4 \pm 1.7 \text{ s}} \tag{5}$$

links both λ and V_{ud} to the average lifetime of the neutron τ_n. By propagating the errors, we find the prediction $\tau_n(\text{SM}) = 878.38 \pm 0.89$ s. Since the neutron decay lifetime τ_n is measured, Eq. (5) could help us to improve the inferences on the IBD cross section. There are two methods of measurement for the neutron decay lifetime. In experiments "in-bottle" or "storage", ultra-cold neutrons are trapped and, after a holding time T, they are released and the number $N(T)$ of surviving ultra-cold neutrons is counted. This is repeated for two or more different holding times, with T ranging from a few minutes up to some fraction of an hour. If no ultra-cold neutrons are lost, τ_n is obtained from an exponential fit to $N(T) \propto \exp(-T/\tau_n)$, and no absolute measurement is required. In "beam" experiments the neutron decay lifetime is directly calculated by counting decay products (protons and/or electrons) from a thermal neutron beam, while simultaneously measuring the neutron beam density.[3] There is a discrepancy among these results. In particular the value reported by the most precise beam experiment conducted at the NIST Center for Neutron Research [14, 15] is 8.9s (3.9σ) higher than the average storage value. This discrepancy between the two set of τ_n measurements has been widely discussed in literature, without reaching general consensus. Since the values of V_{ud} and λ are consistent with the values from the storage approach, and incompatible with the ones from the beam approach, we use only the former data set, assuming that the others are affected by a systematic deviation, which is not yet fully understood. The observed value of the neutron lifetime, together with the SM prediction, yields the relationship $V_{ud} = 2.36323(75)/\sqrt{1+3\lambda^2}$.

> In summary, by propagating the uncertainty factors we find that the cross section is known with $\delta\sigma_{\bar{\nu}_e p}/\sigma_{\bar{\nu}_e p} = 0.1\%$ for low values of electron anti-neutrino energies, which is four times better than [7].

At higher energies, the cross section becomes sensible to the momenta dependence of the form factors. This is generally expressed by phenomenological descriptions as the commonly used dipolar approximation. Since we are mostly concerned with low energy processes, whatever the expression of the full dependence is, we only need the first terms of its Taylor expansion. This is a rather general description and lessens the dependence on the phenomenological approximations, which in the most common cases are not even optimised for the energies we are discussing. Thus we set $g_1/g_1(0) = 1 + q^2 r_A^2/6$. The current value of the axial mass [16] $M_A = 1014 \pm 14$ MeV in the dipolar approximation implies $r_A^2 = 0.455 \pm 0.013$ fm^2; a determination that does not assume the double dipole has an error larger of about one order of magnitude $r_A^2 = 0.46 \pm 0.12$ fm^2.

By including the uncertainty on r_A, we find $\delta\sigma_{\bar{\nu}_e p}/\sigma_{\bar{\nu}_e p} = 1.1\%(E_\nu/50 \text{ MeV})^2$ in the region above $\sim$10 MeV, a value which is 3 times larger than in [7].

---

[3] In the past years a third method for measuring the neutron life-time has been devised, which uses spacecraft-based neutron detectors to count the relative neutron flux as a function of altitude [13]. Its precision is not yet competitive.

While the cross section values are consistent with previous ones, the increase in its uncertainty range motivates attempts to improve the description of form factors at the energies we are considering.

3 Conclusions

The IBD plays an important role in present and future experiments studying neutrinos at low and intermediate energy ranges.

The detection of geo-neutrinos, whose energies extend up to about 2.5 MeV, is realised through IBD. In essentially all oscillation studies at reactors, electron antineutrinos are detected by interactions with free protons via IBD. Reactor neutrino energy spectra end at $\sim$10 MeV. Some of supernova experiments based on water Cherenkov detector are primarily sensitive to IBD interactions, revealing the Cherenkov light of the final-state positron. The inverse beta decay of supernova neutrinos can also be observed in scintillation detectors. Neutrino fluxes from supernovae go up to 50 MeV.

A reliable knowledge of the cross section of the IBD is based on a set of theoretical concepts and on measurements of the key parameters. We assessed the uncertainty on the cross section, which is relevant to experimental advances and increasingly large statistical samples. In order to improve the actual estimates, we need to address the reason of discrepancy in τ_n measurements and the unitarity issue in $|V_{ud}|$. This is particularly significant at the low energies which are relevant e.g. for reactor neutrinos. A priority, especially for supernova studies, is also to refine the description of the axial form factor in the 100 MeV range, thus decreasing the uncertainty due to r_A.

Acknowledgments This work was partially supported by INFN research initiative ENP and by the research grant number 2022E2J4RK "PANTHEON: Perspectives in Astroparticle and Neutrino THEory with Old and New messengers" under the program PRIN 2022 funded by the Italian Ministero dell'Università e della Ricerca (MUR) and by the European Union – Next Generation EU.

References

1. H.A. Bethe, R. Peierls, The 'neutrino'. Nature **133**, 532 (1934)
2. C.L. Cowan, F. Reines et al., Detection of the free neutrino: a confirmation. Science **124**, 103 (1956)
3. G. Ricciardi, N. Vignaroli, F. Vissani, A discussion of the cross section $\bar{\nu}_e + p \rightarrow e^+ + n$. [arXiv:2311.16730 [hep-ph]]
4. G. Ricciardi, N. Vignaroli, F. Vissani, An accurate evaluation of electron (anti-)neutrino scattering on nucleons. J. High Energy Phys. **08**, 212 (2022)
5. A. Kurylov, M.J. Ramsey-Musolf, P. Vogel, Radiative corrections in neutrino deuterium disintegration. Phys. Rev. C **65**, 055501 (2002)

6. P. Vogel, J.F. Beacom, Angular distribution of neutron inverse beta decay, $\bar{v}_e + p \to e^+ + n$. Phys. Rev. D **60**, 053003 (1999)
7. A. Strumia, F. Vissani, Precise quasielastic neutrino nucleon cross-section. Phys. Lett. B **564**, 42 (2003)
8. S. Weinberg, Charge symmetry of weak interactions. Phys. Rev. **112**, 1375–1379 (1958)
9. J.C. Hardy, I.S. Towner, Superallowed $0^+ \to 0^+$ nuclear β decays: 2020 critical survey, with implications for V_{ud} and CKM unitarity. Phys. Rev. C **102**(4), 045501 (2020)
10. R.L. Workman et al., [Particle data group], Review of particle physics. Prog. Theor. Exp. Phys. **2022**, 083C01 (2022)
11. B. Märkisch et al., Measurement of the weak axial-vector coupling constant in the decay of free neutrons using a pulsed cold neutron beam. Phys. Rev. Lett. **122**(24), 242501 (2019)
12. A. Czarnecki, W.J. Marciano, A. Sirlin, Radiative corrections to neutron and nuclear beta decays revisited. Phys. Rev. D **100**(7), 073008 (2019)
13. J.T. Wilson et al., Measurement of the free neutron lifetime using the neutron spectrometer on NASA's Lunar Prospector mission. Phys. Rev. C **104**(4), 045501 (2021)
14. A.T. Yue et al., Improved determination of the neutron lifetime. Phys. Rev. Lett. **111**(22), 222501 (2013)
15. F.E. Wietfeldt et al., Comments on Systematic Effects in the NIST Beam Neutron Lifetime Experiment
16. A. Bodek, S. Avvakumov, R. Bradford, H.S. Budd, Vector and axial nucleon form factors: a duality constrained parameterization. Eur. Phys. J. C **53**, 349 (2008)

Open Access This chapter is licensed under the terms of the Creative Commons Attribution 4.0 International License (http://creativecommons.org/licenses/by/4.0/), which permits use, sharing, adaptation, distribution and reproduction in any medium or format, as long as you give appropriate credit to the original author(s) and the source, provide a link to the Creative Commons license and indicate if changes were made.

The images or other third party material in this chapter are included in the chapter's Creative Commons license, unless indicated otherwise in a credit line to the material. If material is not included in the chapter's Creative Commons license and your intended use is not permitted by statutory regulation or exceeds the permitted use, you will need to obtain permission directly from the copyright holder.

The KM3NeT Underwater Neutrino Telescope

Immacolata Carmen Rea

1 Introduction

Neutrinos are elusive particles. They can be observed indirectly, by detecting the Cherenkov radiation emitted along the path of the secondary charged particles produced after their weak interaction with the Earth or seawater.

As neutrinos interact very rarely, huge targets are needed to increase their probability of interaction. For this reason, their detection technique is based on instrumenting very large volumes of deep ice or water with a matrix of light detectors.

The atmospheric muon background, generated mainly by the interaction of cosmic rays with the atmosphere, decreases at increasing detector depth but to reliably distinguish the neutrino events from atmospheric cosmic rays-induced particles, neutrino telescopes look downwards at upgoing tracks, using the Earth as a shield. In this way, the atmospheric muon background is completely cut out.

I. C. Rea on behalf of KM3NeT Collaboration.

I. C. Rea (✉)
INFN Sez. di Napoli, Complesso universitario di Monte S. Angelo ed. 6 via Cintia, Napoli, Italy
e-mail: imma.rea@na.infn.it

2 The KM3NeT Neutrino Telescope Design: ARCA and ORCA Technologies

The general design of a neutrino telescope consists of arrays with vertical string-like structures coupled to electro-optical cables and equipped with optical modules for the Cherenkov light detection. These detection units have typical lengths of several hundreds of meters and can be distributed in specific geometries according to the energy region of study. The optical modules can host one or more photomultipliers (PMT) with different photocathode areas.

The KM3NeT neutrino telescopes are currently under construction at two different sites in the Mediterranean Sea: ORCA is in France, while ARCA is in Italy. The two detectors share the same technology but have different geometries as they target different energy regions. KM3NeT/ARCA will instrument 1 km^3 of water with 230 detection units (DU) and will detect high-energy neutrinos mainly in the TeV-PeV region. KM3NeT/ORCA will instrument 7 megatonnes of water with 115 DUs and will be dedicated to low-energy studies in the MeV-GeV region. In this way, with the entire KM3NeT programme, it will be possible to span from supernova explosion, and neutrino oscillation studies, up to high-energy astrophysical neutrino detection. The KM3NeT technology is based on a modular design, as shown in Fig. 1. The detection unit consists of a mooring line 800 m long for ARCA and 400 m for ORCA, equipped with 18 digital optical modules (DOM) [1]. The main difference between the two detectors is the density of photosensors: the DOMs are spaced by 36 m along an ARCA-DU while are 9 m apart in an ORCA-DU. The distance between ARCA strings is 90 m, while the ORCA string interspace is 20 m.

The digital optical module has 31 PMTs (with a 3" photocathode) mounted inside a 17" diameter high-pressure-resistant glass sphere. Besides the electronics, inside several calibration devices are hosted. All signals collected are digitised and sent

Fig. 1 KM3NeT detection unit

to shore, where a farm of processors reduces the data volume and filters the signals from the background [1]. Because it is located in the northern hemisphere, KM3NeT has a complementary field of view to IceCube and it points towards the southern sky. One can say that its geometry is "optimised for the Milky Way," as the visibility of the Galactic center and the Galactic plane is about 70%.

3 The KM3NeT Performances: The Effective Area and the Angular Resolution

In general, weak neutrino interactions generate different event topologies depending on the flavour of the neutrino and on the charged current (CC) or neutral current (NC) interaction. This can lead to upgoing track-like events generated by CC-interactions of muon neutrinos that can travel for a long path within the detector, or to shower-like events related to CC-interactions of electron neutrinos.

An evaluation of the *effective areas* (A_{eff}) for the different partial configurations of the ARCA detectors for a flux of muon neutrinos and muon anti-neutrinos that interact via CC interaction can be seen in Fig. 2 (on the left) [2]. Compared to the ANTARES one for similar upgoing event selection, the A_{eff} for ARCA6 and ARCA8 was lower up to 10^5 GeV; with the new configuration of ARCA19/21 the A_{eff} increased, becoming comparable to the ANTARES one already at low energies. The acronyms ARCA 6, ARCA8, ARCA19, and ARCA21 indicate the ARCA configurations with 6, 8, 19 and 21 detection units respectively. In Fig. 2 on the right the *angular resolution* of the telescope is reported [2]: it is not trivial to obtain a significant pointing accuracy to identify astrophysical sources of neutrinos. Nevertheless, the period with ARCA6-8 reached an angular resolution of 1° at $100*10^{15}$ eV, while in the ARCA21 configuration, the resolution for track-like events improved up to 0.2°. This resolution is expected to improve further for the full detector reaching the value of 0.06°.

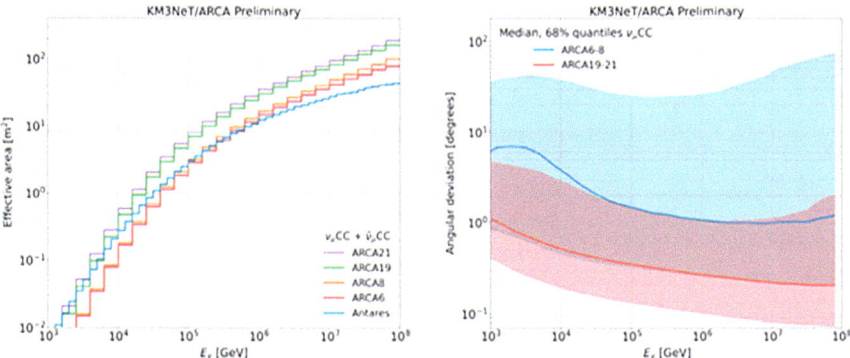

Fig. 2 Left: Effective areas for the different ARCA detectors compared to ANTARES. **Right:** Angular resolution [2]. © 2023, I. C. Rea on behalf of the KM3NeT Collaboration

4 KM3NeT/ARCA: First Results

The search for point-like sources is among the primary science goals of KM3NeT. For the ARCA detector, a series of upper limits on neutrino emission flux at different configurations have been estimated [2]: in Fig. 3 the sensitivities that have been calculated for ARCA6, ARCA6-8 and ARCA6-21 at increasing observation periods (respectively 92, 302 and 424 days of data taking) are reported. These 3 measurements are compared in the same plot with the full final ARCA performance for 10 years of data taking, and with 15 and 10 years of ANTARES and IceCube. The sensitivity of ANTARES will be approached very soon thanks to the next upgrades and, when the entire ARCA detector will be completed, it will reach a sensitivity level comparable to the IceCube one for the northern hemisphere and will improve it by almost a factor 50 for the Southern hemisphere.

The detection of the neutrino diffuse flux is a key factor for the understanding of the production and acceleration mechanisms of cosmic rays. In addition, it can open a new observation window on faint sources that otherwise would be lost. In the plot in Fig. 4, the upper limits to a diffuse neutrino flux at the early stages of the ARCA detector are shown [3]: after 432 days of data-taking, a significant improvement has been achieved with the ARCA21 configuration, comparable with the ANTARES and the IceCube performances.

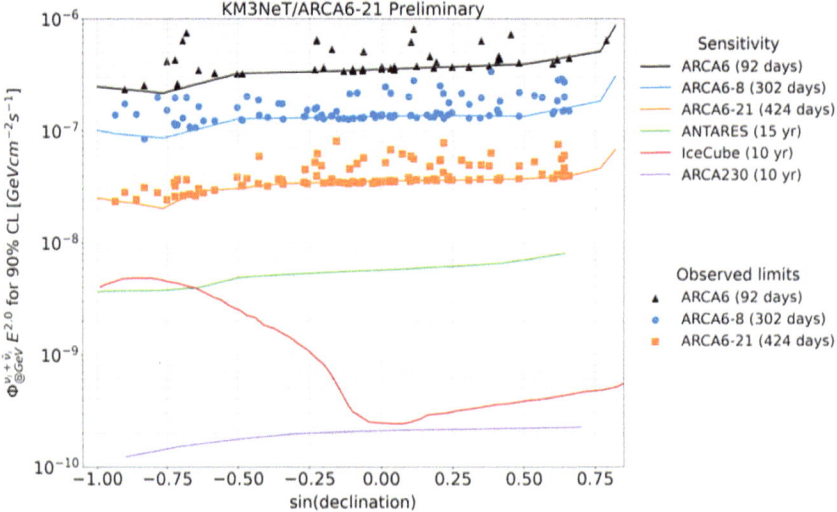

Fig. 3 Comparison of the observed limits on the flux for the ARCA6-21 point source sensitivity. Reprinted under CC-BY-NC-ND-4.0 from [2]. © 2023, I. C. Rea on behalf of the KM3NeT Collaboration

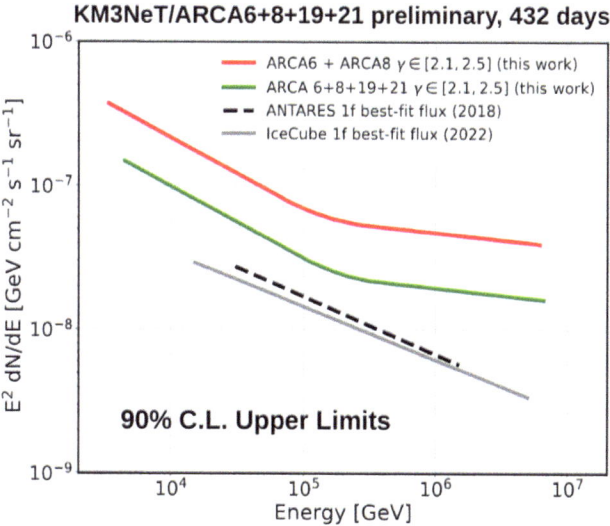

Fig. 4 90% C.L. upper limits to a diffuse neutrino flux at the early stages of the detector construction. Reprinted under CC-BY-NC-ND-4.0 from [3]. © 2023, I. C. Rea on behalf of the KM3NeT Collaboration

5 KM3NeT/ORCA: First Results

The ORCA detector is optimised in the low-energy range (MeV-GeV) and it aims at determining the neutrino mass hierarchy, studying the neutrino oscillation patterns, by detecting the neutrino flux generated in the Earth's atmosphere. Data collected with the ORCA6 configuration confirmed that the oscillation scenario is preferred over the hypothesis of no oscillation [4]. In Fig. 5 the contour at 90% of C.L. for the oscillation parameters $sin^2\theta_{23}$ and Δm_{31}^2, constrained simultaneously and compared with other experiments, is shown. These preliminary measurements estimated as the best fit values $sin^2\theta_{23} = 0.51^{+0.06}_{-0.07}$ and $\Delta^2 m_{31} = 2.14^{+0.36}_{-0.25} * 10^{-3}$ eV^2 with a mild preference for normal mass ordering $-2log(L_{NO}/L_{IO}) = 0.9$. Currently, ORCA is at 5% of its final configuration and these measurements will gain precision as the detector volume increases.

6 Conclusions

The KM3NeT network of neutrino telescopes, ARCA and ORCA, is under construction. Both detectors have started to collect data in their partial setup. In this contribution, preliminary results have been presented and discussed together with expected performances of the final configuration.

Fig. 5 Contour at 90% C.L. of ORCA6 for the oscillation parameters $sin^2\theta_{23}$ and Δm_{31}^2 constrained simultaneously and compared with other experiments. Reprinted under CC-BY-NC-ND-4.0 from [4]. © 2023, I. C. Rea on behalf of the KM3NeT Collaboration

References

1. S. Aiello et al., The KM3NeT multi-PMT optical module. J. Instrum. **17**, P07038 (2022)
2. R. Muller et al., Search for cosmic neutrino point sources and extended sources with 6-21 lines of KM3NeT/ARCA. Proc. Sci. **ICRC2023**, 1018 (2023)
3. V. Tsourapis et al., Search for a diffuse astrophysical neutrino flux with KM3NeT/ARCA using data of 2021–2022. Proc. Sci. **ICRC2023**, 1195 (2023)
4. V. Carretero et al., Measuring atmospheric neutrino oscillation with KM3NeT/ORCA6. Proc. Sci. **ICRC2023**, 996 (2023)

Open Access This chapter is licensed under the terms of the Creative Commons Attribution 4.0 International License (http://creativecommons.org/licenses/by/4.0/), which permits use, sharing, adaptation, distribution and reproduction in any medium or format, as long as you give appropriate credit to the original author(s) and the source, provide a link to the Creative Commons license and indicate if changes were made.

The images or other third party material in this chapter are included in the chapter's Creative Commons license, unless indicated otherwise in a credit line to the material. If material is not included in the chapter's Creative Commons license and your intended use is not permitted by statutory regulation or exceeds the permitted use, you will need to obtain permission directly from the copyright holder.

SATURNE: The Sarov Tritium Neutrino Experiment for Probing Coherent Elastic Neutrino-Atom Scattering and Neutrino Electromagnetic Interactions

M. Cadeddu, F. Dordei, C. Giunti, A. P. Ivashkin, K. A. Kouzakov, F. M. Lazarev, O. A. Moskalev, I. S. Stepantsov, A. I. Studenikin, I. I. Tkachev, V. N. Trofimov, M. A. Verkhovtsev, M. M. Vyalkov, A. A. Yukhimchuk, and E. F. Zagirdinova

1 Introduction

In the Standard Model (SM), the neutrinos are massless particles coupled only to the Z^0 and $W^{\pm}$ gauge bosons. The established existence of non-zero neutrino masses opens up the possibility for neutrinos to have other properties that are beyond the SM. Of particular interest in this regard is the search for neutrino electromagnetic characteristics, such as the electric charge (millicharge), charge radius, magnetic,

---

This research was supported by the Russian Science Foundation (project # 22-22-00384).

M. Cadeddu · F. Dordei
Istituto Nazionale di Fisica Nucleare (INFN), Sezione di Cagliari, Complesso Universitario di Monserrato S.P. per Sestu Km 0.700, Monserrato (Cagliari), Italy

C. Giunti
Istituto Nazionale di Fisica Nucleare (INFN), Sezione di Torino, Torino, Italy

A. P. Ivashkin · I. I. Tkachev
Institute for Nuclear Research, Russian Academy of Sciences, Moscow, Russia

K. A. Kouzakov (✉) · F. M. Lazarev · I. S. Stepantsov · A. I. Studenikin · E. F. Zagirdinova
Faculty of Physics, Lomonosov Moscow State University, Moscow, Russia

O. A. Moskalev · A. A. Yukhimchuk
Russian Federal Nuclear Center – All-Russian Scientific Research Institute of Experimental Physics, Sarov, Nizhny Novgorod Region, Russia

V. N. Trofimov
Dzhelepov Laboratory of Nuclear Problems, Joint Institute for Nuclear Research, Dubna, Moscow Region, Russia

electric and anapole moments. The discovery of the indicated neutrino properties will literally open a window to new physics [1].

So far, the most studied neutrino electromagnetic characteristic, both theoretically and experimentally, is the neutrino magnetic moment. A nonzero magnetic moment for neutrinos arises already in the minimal SM extension with addition of the right-handed Dirac massive neutrinos. The predicted value is [2]

$$\mu_\nu = \frac{3\sqrt{\alpha} G_F m_\nu}{8\sqrt{2}\pi^2} \approx 3.2 \times 10^{-19} \left(\frac{m_\nu}{1\,\text{eV}}\right) \mu_B, \qquad (1)$$

where G_F is the Fermi constant, α is the fine-structure constant, m_ν is the neutrino mass, and μ_B is the Bohr magneton. There are various theoretical scenarios beyond the minimally extended SM which predict much greater neutrino magnetic moment values compared to (1) (see the review article [1] and references therein). Moreover, some of such theoretical predictions have been already ruled out by the world leading experimental constraints on the neutrino magnetic moment. Such constraints are obtained in laboratory experiments and astrophysics (see [3] and references therein).

SATURNE will search for neutrino electromagnetic interactions in elastic and ionizing neutrino-atom collisions. A high-intensity tritium neutrino source will be employed in combination with the He-4, Si and SrI_2(Eu) targets in order to study the elastic and ionization channels of neutrino-atom collisions at unprecedentedly low energies. With studying the ionization of atoms by tritium neutrinos, SATURNE has a potential to improve the strongest laboratory limits [4] to a level that makes them competitive or even better than the world-leading upper limits [3]. It has an even greater potential with studying coherent elastic neutrino-atom scattering (CEνAS) [5, 6]. In such a case, the sensitivity to the μ_ν value can be almost an order of magnitude better than the world-leading upper limits.

2 Overview of SATURNE

SATURNE is a part of the scientific program of the National Center for Physics and Mathematics (NCPhM) that was founded in Sarov in 2021. The experiment is under preparation, with the first taking of data anticipated to begin in 2025 and the data

---

M. A. Verkhovtsev
Branch of Lomonosov Moscow State University in Sarov, Sarov, Nizhny Novgorod Region, Russia

M. M. Vyalkov
Russian Federal Nuclear Center – All-Russian Scientific Research Institute of Experimental Physics, Sarov, Nizhny Novgorod Region, Russia
Branch of Lomonosov Moscow State University in Sarov, Sarov, Nizhny Novgorod Region, Russia

collection expected to be completed by 2032. A marked feature of the experiment is the use of a high-intensity tritium neutrino source, with initial activity of at least 10 MCi and possibly up to 40 MCi. The Si and SrI$_2$(Eu) detectors with record low-energy thresholds for such detector types will measure the ionization channel of neutrino-atom collisions, namely

$$\bar{\nu}_e + \mathcal{A} \rightarrow \mathcal{A}^+ + e^- + \bar{\nu}_\ell, \qquad (2)$$

where $\mathcal{A}$ is the atomic target, and the final antineutrino flavor ℓ may differ from the electron flavor if neutrino electromagnetic interactions are present. In its turn, the measurements with the liquid He-4 detector in a superfluid state are expected to provide the first observation of the CEνAS process

$$\bar{\nu}_e + \mathcal{A} \rightarrow \mathcal{A} + \bar{\nu}_\ell. \qquad (3)$$

This will bring the experimental studies of coherent elastic neutrino-nucleus scattering (CEνNS) [7] to a qualitatively new level, namely when one will be able to explore the neutrino elastic scattering not only on a nucleus as a whole, but also on an atom as a whole.

Low-Background Neutrino Laboratory For conducting the SATURNE measurements, a low-background neutrino laboratory is being created at the NCPhM. For this purpose an existing shallow underground facility in Sarov is used. It is equipped with a service lift, waterproofed and meets the necessary requirements for working with a high-intensity tritium source. The facility has an overburden consisting of 4-m thick reinforced concrete, manufactured before the Atomic Age and in this regard pure in terms of uranium series elements, 5-m thick sand and a layer of soil, which is more than 0.5-m thick. The overburden is about 20-25 m of water equivalent, which is enough to sufficiently suppress the soft and hadronic components of cosmic radiation for acceptable background conditions.

Tritium Neutrino Source Tritium is an unstable hydrogen isotope with a half-life of 12.3 years. The tritium β^- decay,

$$^3\text{H} \rightarrow {}^3\text{He} + e^- + \bar{\nu}_e, \qquad (4)$$

produces electron antineutrinos in the energy range from 0 to 18.6 keV, with an average energy of about 13 keV. A tritium neutrino source is thus well suited for studying CEνAS that requires neutrino energies $E_\nu \lesssim 10$ keV [5].

The principal design of a tritium neutrino source was worked out in [8]: it is a set of tubular elements, in which tritium is in a chemically bound state on titanium. Titanium powder in bulk form is placed in the tubular element. Then the titanium powder is thermally activated and saturated with tritium, after which the tubular element is sealed. This tubular element design will be classified as a "closed radionuclide source": the design of the tubular element is made in such a way that it is possible to extract tritium from it only under special factory conditions.

Si Crystal Detector The Si detector is supposed to consist of fourteen 125-cm^3 Si crystal cylinders with a total mass of 4 kg, operating at a temperature between 10 and 50 mK, surrounded with the tubular elements of the tritium neutrino source. The low operating temperature is designed to take advantage of the Neganov–Trofimov–Luke effect [9, 10], that is, when the conversion of ionization energy into heat is amplified by a (high) bias voltage. The heat is measured by microcalorimeters, for example, such as transition edge sensors (TESs) [11], which are mounted on each silicon cylinder. This may allow to achieve the energy threshold of the detector as low as about 10 eV or even 1 eV [12]. In such a case, after 1 year of collecting data with the Si detector using 1 kg of tritium, one can achieve the sensitivity to the $\mu_{\bar{\nu}_e}$ value on the order of $1.6 \times 10^{-12} \mu_B$ at 90% C.L.

SrI$_2$(Eu) Scintillation Detector The scintillation detector is assumed to be an assembly of SrI$_2$(Eu) crystal elements with a total mass of 14 kg surrounding the tritium neutrino source. The elements, each having dimensions of $15 \times 15 \times 50$ mm^3 and a mass of 50 g, are combined into modules with four elements per module. The operating temperature of the detector is in the range of 80–100 K. At such a temperature the thermal noise of silicon photomultipliers used for signal readout decreases sharply, thus making it possible to detect extremely low light signals (at the level of several photons only). This is expected to provide a record light collection and a detector threshold as low as 100 eV. In the latter case, after 1 year of collecting data with the SrI$_2$(Eu) detector using 1 kg of tritium, one can achieve the sensitivity to the $\mu_{\bar{\nu}_e}$ value on the order of $2 \times 10^{-12} \mu_B$ at 90% C.L.

Liquid He-4 Detector The concept of the liquid He-4 detector assumes that the tritium neutrino source is placed inside the hollow of the cylinder tank filled with a cubic meter of liquid He-4 in a superfluid state. The dilution refrigerator keeps liquid He-4 at a temperature between 40 and 60 mK. An array of about 1000 TES microcalorimeters suspended in vacuum above the helium surface registers the evaporated He-4 atoms, which are induced by scattering of electron antineutrinos from the tritium β-decay on atoms in the superfluid. The signal from the TES microcalorimeters is readout by a SQUID system. In such a way, the He-4 detector is expected to be sensitive to the energy signals below 0.1 eV. After 5 years of collecting data with the He-4 detector, one can achieve the sensitivity to the $\mu_{\bar{\nu}_e}$ value as high as $(2-3) \times 10^{-13} \mu_B$ at 90% C.L., depending on the initial activity of the tritium source.

3 Summary and Outlook

The tritium neutrino experiment, SATURNE, is under preparation at the NCPhM in Sarov. Its major goal is the first observation of coherent elastic neutrino-atom scattering, in order to test SM neutrino interactions at unprecedentedly low energies and to search for new physics, in particular, such as the neutrino magnetic moment. For these purposes, a low-background neutrino laboratory is being created and a high-intensity tritium neutrino source is being prepared with a tritium mass of at least 1 kg (10 MCi), and possibly up to 4 kg (40 MCi).

A 1-m^3 superfluid helium-4 detector is being developed for the first observation of coherent elastic neutrino-atom scattering. After 5 years of collecting data, by 2032, it is expected to achieve the neutrino magnetic moment sensitivity which is an order of magnitude better than the current world leading constraints [3, 4].

A 14-kg SrI$_2$(Eu) scintillation detector and a 4-kg Si crystal detector are being developed for studying atomic ionization by tritium neutrinos. After 1 year of collecting data, by 2026 and 2027, respectively, they are expected to achieve the neutrino magnetic moment sensitivity that is competitive with or even better than the current world leading constraints.

Acknowledgments We are grateful to Andrey Pankratov, Anna Gordeeva, Alexander Mel'nikov and Vladimir Eremein for useful discussions. The work on the Si and SrI$_2$(Eu) detectors is conducted within the scientific program of the National Center for Physics and Mathematics (Section No. 8, Stage 2023-2025). The study on the He-4 detector is supported by the Russian Science Foundation (project #22-22-00384).

References

1. C. Giunti, A. Studenikin, Rev. Mod. Phys. **87**(2), 531 (2015)
2. K. Fujikawa, R.E. Shrock, Phys. Rev. Lett. **45**(12), 963 (1980)
3. S. Navas et al. (Particle Data Group), Phys. Rev. D **110**(3), 030001 (2024)
4. A. Khan, Phys. Lett. B **837**, 137650 (2023)
5. Yu.V. Gaponov, V.N. Tikhonov, Sov. J. Nucl. Phys. **26**(3), 314 (1977)
6. M. Cadeddu, F. Dordei, C. Giunti, K.A. Kouzakov, E. Picciau, A.I. Studenikin, Phys. Rev. D **100**(7), 073014 (2019)
7. V. Pandey, Prog. Part. Nucl. Phys. **134**, 104078 (2024)
8. A.A. Yukhimchuk et al., Fusion Sci. Technol. **48**(1), 294 (2005)
9. B. Neganov, V. Trofimov, Otkrytia i Izobreteniya **146**, 215 (1985) (in Russian)
10. P.N. Luke, J. Appl. Phys. **64**(12), 6858 (1988)
11. V.Y. Safonova et al., Materials **17**(1), 222 (2024)
12. B.S. Neganov, V.N. Trofimov, A.A. Yukhimchuk, L.N. Bogdanova, A.G. Beda, A.S. Starostin, Phys. At. Nucl. **64**(11), 1948 (2001)

Open Access This chapter is licensed under the terms of the Creative Commons Attribution 4.0 International License (http://creativecommons.org/licenses/by/4.0/), which permits use, sharing, adaptation, distribution and reproduction in any medium or format, as long as you give appropriate credit to the original author(s) and the source, provide a link to the Creative Commons license and indicate if changes were made.

The images or other third party material in this chapter are included in the chapter's Creative Commons license, unless indicated otherwise in a credit line to the material. If material is not included in the chapter's Creative Commons license and your intended use is not permitted by statutory regulation or exceeds the permitted use, you will need to obtain permission directly from the copyright holder.

An Overview of the DUNE Far Detectors

Leïla Haegel

1 The DUNE Physics Program

The Deep Underground Neutrino Experiment (DUNE) is a long-baseline neutrino oscillation experiment aiming at observing the transition from muon (anti)-neutrinos to electron and tau (anti)-neutrinos [1]. A $\stackrel{(-)}{\nu_\mu}$ broadband beam peaking at 2.5 GeV will be created in Fermilab PIP-II accelerator and sent towards the far detector complex, located 1300 km away in the Sanford Underground Research Facility (SURF). The far detector complex will hold four ∼20-kton far detectors (FDs) based on liquid-argon technology in order to observe oscillation towards $\stackrel{(-)}{\nu_e}$ and $\stackrel{(-)}{\nu_\tau}$ flavours. In Phase I of the experiment, the complex will hold two detectors (FD1 and FD2), as well as a moveable liquid argon near detector located downstream the Fermilab beam dump in order to characterise the neutrino flux. In Phase II, two detectors (FD3 and FD4) will be added in the far detector complex, and an improved near detector complex will provide better measurements of the unoscillated neutrino flux.

The physics objectives of DUNE are the measurement of charge-parity (CP) violation in the leptonic sector, with the ability to reach a 5σ precision over 50% of the CP-violating phase δ_{CP} values [2]. Using the difference of scattering in ν_e and $\bar{\nu}_e$ along the long beamline, DUNE also has the capacity to measure the ordering of neutrino mass eigenstates at 5σ as shown on Fig. 1. Concerning the oscillation

---

Leïla Haegel on behalf of the DUNE collaboration.

L. Haegel (✉)
Universite Claude Bernard Lyon 1, CNRS/IN2P3, IP2I Lyon, UMR 5822, Villeurbanne, France
e-mail: l.haegel@ip2i.in2p3.fr

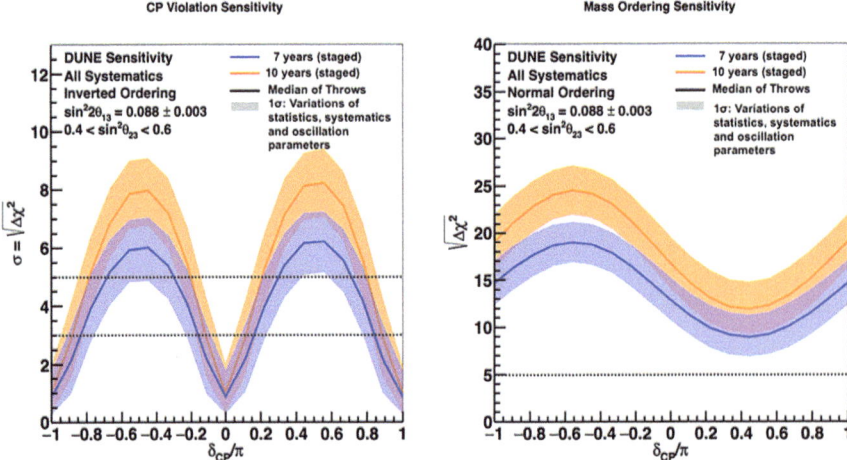

Fig. 1 DUNE sensitivity to CP violation (left) and neutrino mass ordering (right) as a function of the CP-violating phase δ_{CP}, assuming the normal ordering of neutrino masses. Reprinted under CC-BY-4.0 from [2]. © 2024, The Author(s)

matrix angle, DUNE can resolve the θ_{23} octant in case of non-maximal mixing and provide an independent measurement of θ_{13} with a resolution comparable to reactor experiments at high exposure. In addition to beam neutrinos, DUNE can also observe neutrinos at lower energies originating from the Sun or supernovae, and neutrinos at higher energies created during the collision of cosmic rays with the Earth atmosphere. Besides its precision neutrino oscillation physics measurements program, DUNE will be able to probe a large range of beyond-Standard Model phenomena, from nucleon decay to spacetime symmetry breaking and new particles and interactions [3].

2 The DUNE Far Detector Complex

The scientific program described above requires a large number of neutrino observations, a known challenge considering the very low cross-section of those particles. For this aim, in Phase I, the DUNE far detector complex will include two far detectors placed in $66 \times 19 \times 18$ m^3 cryostats in order to observe $\mathcal{O}(10^3)$ neutrino interactions over the experiment timeline. The cryostats are subdivised in modules filled with liquid argon at a temperature of 88 K ($-185\,°$C) in order to detect the charged particles created by neutrino interactions in the 10-kton fiducial volume. At DUNE energies, neutrinos interact with argon atoms predominantly through charge-current resonant, quasi-elastic and deep inelastic interactions creating charged particles. In 79% of the cases, the argon atom is ionised and the created electrons are drifted towards an anode acting a signal collector by a 500-kV electric field. In 21% of the cases, the atom is excited and form an excimer state with another

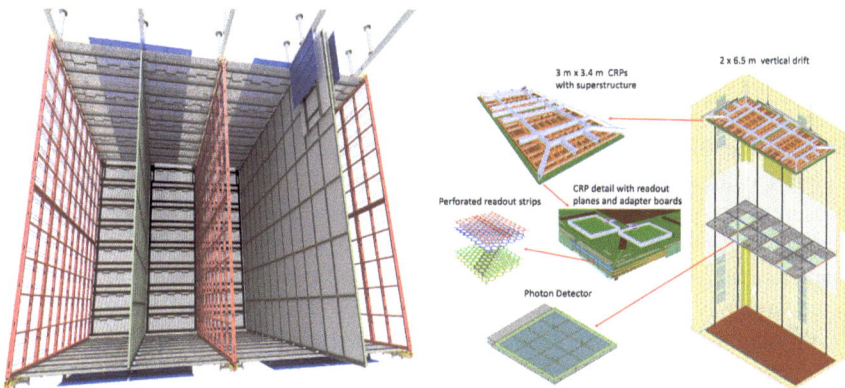

Fig. 2 Left: schematics of a FD1 (horizontal drift) module, where the red planes are the wire plane anodes and the grey planes the cathodes (from [4]). Right: schematics of a FD2 (vertical drift) module, where the top and bottom planes are the CRP anodes, and the middle plane is the cathode. Reprinted under CC-BY-4.0 from [5]. © 2024, The Author(s)

atom, releasing the extra energy through ultra-violet photons collected by an X-ARAPUCA light trap made of wavelength shifting, dichroic filer and silicon phototmultiplier.

While both FD1 and FD2 rely on similar detection principle, they present small variations in their design. The FD1 modules contain 3 anodes and 2 cathodes placed vertically, creating four drift volumes of 3.6 m length as shown on Fig. 2 left. The charge signal is collected at the anode plane assembly, consisting of 3 layers of wire planes, each plane holding 2560 wires with a 4.7 mm pitch [4]. Each anode plane contains 10 light traps, and the read-out is performed by cryogenic electronics located inside the cryostat. The FD2 modules contain 2 anodes and 1 cathode placed horizontally, creating a vertical electric field with 6.5 m drift length as shown on Fig. 2 right. The ionisation electrons are collected at the anode consisting of 160 charge readout planes across the detector [5]. Each plane is made of 3 layers of circuit board printed with linear copper strips and drilled with holes to enable the circulation of charges, with a total of 3072 strips per plane with a pitch from 5 to 7.65 mm. The bottom electronics is similar to FD1, while the top electronic is divided between the amplification stage inside the cryostat and the digitisation stage located outside for better accessibility and longevity. The light collection is assured by 352 X-ARAPUCA located on the cathods and in the field cage.

3 Current Status and Future Prospects

The DUNE collaboration carried a large R&D program over the last years. Small prototypes have been extensively used to test the design of the FD1 and FD2 cathodes and anodes. A first module of the FD1, labelled ProtoDUNE-SP, of volume $7 \times 6 \times 7.2$ m, have been assembled and put to test in the CERN Neutrino platform between 2018 and 2020 [6]. Using cosmic rays and the H4 beamline delivering charged pions, kaons, protons, muons and electrons with momenta in the range 0.3–7 GeV/c, the detector performance has been validated within design specifications, and exceeding them in several cases [7]. The large set of measurements performed includes the energy loss for charged particle, the drift electron lifetime, the electronic gain and noise, as well as the light detectors characteristics. The identification and reconstruction of particles going through liquid argon is performed using the LArSoft toolkit developed and maintained for the ArgoNeuT, DUNE, LArIAT, MicroBooNE, and SBND experiments[1] [8], and furthermore calibrated using ProtoDUNE data to achieve the required sub-centimeter resolution. ProtoDUNE-SP, relabelled ProtoDUNE-HD, will take more data in Summer 2024. A first module of the FD2, ProtoDUNE-VD, is currently being assembled and will take data in 2025. The far detectors are planned to be installed and start taking by the end of the 2020s.

Acknowledgments This document was prepared by the DUNE collaboration using the resources of the Fermi National Accelerator Laboratory (Fermilab), a U.S. Department of Energy, Office of Science, HEP User Facility. Fermilab is managed by Fermi Research Alliance, LLC (FRA), acting under Contract No. DE-AC02-07CH11359. The ProtoDUNE detector was constructed and operated on the CERN Neutrino Platform. We gratefully acknowledge the support of the CERN management, and the CERN EP, BE, TE, EN and IT Departments for NP04/ProtoDUNE-SP. This work was supported by CNPq, FAPERJ, FAPEG and FAPESP, Brazil; CFI, IPP and NSERC, Canada; CERN; MŠMT, Czech Republic; ERDF, H2020-EU and MSCA, European Union; CNRS/IN2P3 and CEA, France; INFN, Italy; FCT, Portugal; NRF, South Korea; CAM, Fundación "La Caixa", Junta de Andalucía FEDER, MICINN, and Xunta de Galicia, Spain; SERI and SNSF, Switzerland; TÜBİTAK, Turkey; The Royal Society and UKRI/STFC, United Kingdom; DOE and NSF, United States of America.

References

1. B. Abi et al. (DUNE Collaboration), Deep Underground Neutrino Experiment (DUNE), far detector technical design report, volume I: introduction to DUNE. J. Instrum. **15**, T08008 (2020). https://doi.org/10.1088/1748-0221/15/08/T08008. [arXiv:2002.02967 [physics.ins-det]]
2. B. Abi et al. (DUNE Collaboration), Long-baseline neutrino oscillation physics potential of the DUNE experiment. Eur. Phys. J. C **80**, 978 (2020). https://doi.org/10.1140/epjc/s10052-020-08456-z. [arXiv:2006.16043 [hep-ex]]
3. B. Abi et al. (DUNE Collaboration), Prospects for beyond the standard model physics searches at the deep underground neutrino experiment. Eur. Phys. J. C **81**, 322 (2021). https://doi.org/10.1140/epjc/s10052-021-09007-w. [arXiv:2008.12769 [hep-ex]]
4. B. Abi et al. (DUNE Collaboration), Deep Underground Neutrino Experiment (DUNE), far detector technical design report, volume IV: far detector single-phase technology. J. Instrum. **15**, T08010 (2020). https://doi.org/10.1088/1748-0221/15/08/T08010. [arXiv:2002.03010 [physics.ins-det]]

---

[1] https://larsoft.org/

5. A. Abed Abud et al. (DUNE Collaboration), The DUNE far detector vertical drift technology, Technical Design Report. [arXiv:2312.03130 [hep-ex]]
6. A.A. Abud, et al. (DUNE Collaboration), Design, construction and operation of the ProtoDUNE-SP Liquid Argon TPC. J. Instrum. **17**, P01005 (2022). https://doi.org/10.1088/1748-0221/17/01/P01005. [arXiv:2108.01902 [physics.ins-det]]
7. B. Abi et al. (DUNE Collaboration), First results on ProtoDUNE-SP liquid argon time projection chamber performance from a beam test at the CERN Neutrino Platform. J. Instrum. **15**, P12004 (2020). https://doi.org/10.1088/1748-0221/15/12/P12004. [arXiv:2007.06722 [physics.ins-det]]
8. E.L. Snider, G. Petrillo, LArSoft: toolkit for simulation, reconstruction and analysis of liquid argon TPC neutrino detectors. J. Phys. Conf. Ser. **898**, 042057 (2017). https://doi.org/10.1088/1742-6596/898/4/042057

Open Access This chapter is licensed under the terms of the Creative Commons Attribution 4.0 International License (http://creativecommons.org/licenses/by/4.0/), which permits use, sharing, adaptation, distribution and reproduction in any medium or format, as long as you give appropriate credit to the original author(s) and the source, provide a link to the Creative Commons license and indicate if changes were made.

The images or other third party material in this chapter are included in the chapter's Creative Commons license, unless indicated otherwise in a credit line to the material. If material is not included in the chapter's Creative Commons license and your intended use is not permitted by statutory regulation or exceeds the permitted use, you will need to obtain permission directly from the copyright holder.

Exploring Entanglement and Quantum Fisher Information in a System of Two Superconducting Qubits Subjected to Thermal Noise

Mourad Benzahra and Mostafa Mansour

1 Introduction

This study delves into the dynamics of a two-qubit superconducting charge system, an essential component of quantum computing technology. Within the realm of quantum information theory, entanglement emerges as a critical resource, crucial for the development of quantum technologies and the execution of quantum tasks [1]. Various metrics, such as concurrence, negativity, and logarithmic negativity, have been proposed and extensively utilized to quantify entanglement in this domain. In recent years, there has been a burgeoning interest in local quantum Fisher information, a measure of quantum correlations that transcends the boundaries of entanglement and finds applications across the spectrum of quantum information science [2]. This study is centered on the analysis of a two-qubit system achieved by coupling a superconducting environment to a superconducting island through the Josephson junction. Its primary aim is to investigate the dynamic interplay of logarithmic negativity and quantum Fisher information within this setup, with the ultimate goal of elucidating its potential for quantum information processing.

M. Benzahra (✉) · M. Mansour
LHEPCM, Department of Physics, Faculty of Sciences Ain Chock, Hassan II University, Casablanca, Morocco

2 Quantum Correlations and Coherence Quantifiers

2.1 Entanglement Quantified by Logarithmic Negativity

Logarithmic negativity ($\mathcal{LN}$) is widely recognized as a significant identifier of the degree of entanglement for a bipartite system. We consider a two-qubit state ρ, the $\mathcal{LN}$ is calculated as follows [3, 4]:

$$\mathcal{LN}(\rho) = log_2(\sum_l |\beta_l|), \quad (1)$$

where β_l are the eigenvalues of the partially transposed of the density matrix (ρ^{T_Y}).

2.2 Non-classical Correlations Quantified by Local Quantum Fisher Information

The local quantum Fisher information of a bipartite density (F(ρ_{XY})) is determined as [5, 6]

$$F(\rho) = 1 - Max\{\alpha_1, \alpha_2, \alpha_3\}, \quad (2)$$

where $\alpha_{i,i=1,2,3}$ represent the eigenvalues of the 3×3 symmetric matrix $\mathcal{M}$ whose components are defined as follows.

$$\mathcal{M}_{m,n} = \sum_{l,k,\gamma_l+\gamma_k>0} \sum_{m,n=1}^{3} \frac{2\gamma_l \gamma_k}{\gamma_l + \gamma_k} \langle \eta_l | \sigma_m \otimes I_Y | \eta_k \rangle \langle \eta_k | \sigma_n \otimes I_Y | \eta_l \rangle, \quad (3)$$

where $\sigma_{m,n}(m, n = x, y, z)$ are the Pauli matrices, concerning $\gamma_{l,k}$ and $|\eta_{l,k}\rangle$ are defined the state of the system with $\rho = \sum_i \gamma_i |\eta_i\rangle \langle \eta_i|$.

3 Physical Model

In this present work, we consider two boxes, each containing a pair of coopers ($SCBs$), these are identical and coupled by a capacitor. The two identical superconducting qubits are established by interfering the quantum state of the two boxes with each other. The depiction of the dynamics of the two superconducting qubits can be expressed by the following Hamiltonian [7, 8]:

$$\mathcal{H} = -\frac{1}{2}([4E_{C1}(\frac{1}{2} - n_{g1}) + 4E_m(\frac{1}{2} - n_{g2})]\sigma_{z1} + [4E_{C2}(\frac{1}{2} - n_{g2}) \quad (4)$$
$$+ 2E_m(\frac{1}{2} - n_{g1})]\sigma_{z2} + E_{J1}\sigma_{x2} + E_{J2}\sigma_{x2} - E_m\sigma_{zz}),$$

where n_{g1} is the qubit gate charge. E_{C1} and E_{J1} are the charging energy and the Josephson energy, while the parameter E_{m1} represents the mutual coupling energy between the qubits under study.

The thermal density matrix of the two superconducting qubits system can be expressed using the Gibbs relation:

$$\rho(T) = \frac{e^{-\beta \mathcal{H}}}{Z}, \quad (5)$$

Here, $\beta = \frac{1}{k_B T}$ is assumed, with the Boltzmann constant set to $k_B = 1$. The partition function Z is then defined as $Z = \text{Tr}(e^{\frac{-\mathcal{H}}{T}})$.

4 Results

In this section, we illustrate the results derived from modelling the dynamics of the two measures on two superconducting qubits system. We investigate the influence of the temperature T, the Josephson energies E_{J12}, and coupling energy E_m on these measures in the density matrix $\rho(T)$. In Fig. 1, we set $E_{J1} = E_{J2} = 1.5$ and select E_m at values of $0, 0.3, 0.5, 1$ and 1.5.

It can be observed from this figure that, when $E_m = 0$, the quantum entanglement and correlations captured by $\mathcal{LN}(\rho)$ and $F(\rho)$ are zero at any temperature. This means that the two-qubit system is unentangled and uncorrelated in the absence of coupling energy. At $T = 0$, increasing the parameter E_m lead to an augmentation of the values of the two measures $\mathcal{LN}(\rho)$ and $LQFI$. Specifically, the function

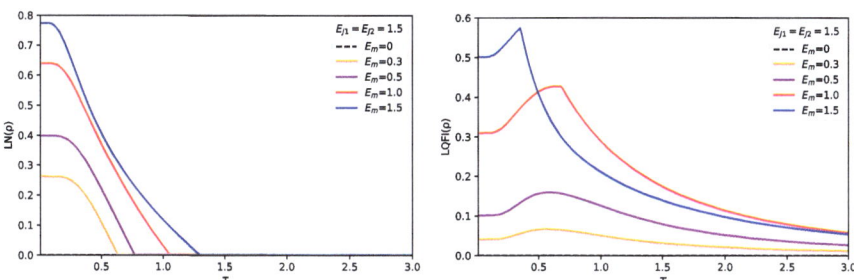

Fig. 1 The behavior of $\mathcal{LN}(\rho)$ and $F(\rho)$ as function of T for various values of E_m, with $E_{J1} = E_{J2} = 1.5$

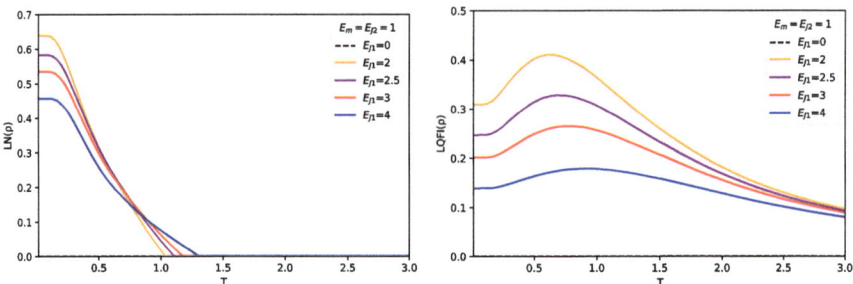

Fig. 2 The behavior of $\mathcal{LN}(\rho)$ and $F(\rho)$ as function of T for various values of E_{J1}, with $E_m = E_{J2} = 1$

$\mathcal{LN}(\rho)$ attends its maximum at this temperature. Now, as the temperature increases, $\mathcal{LN}(\rho)$ continues to decrease until it vanishes at a specific temperature value. Moreover, for high values of E_m, it is noticeable that $\mathcal{LN}(\rho)$ and $F(\rho)$ become more resistant to the effects of temperature. On the other hand, we can see that $LQFI$ exhibits a different behavior compared to $\mathcal{LN}(\rho)$. Indeed, at low temperatures, the quantifier $LQFI$ increases with T until it reaches its maximum value and subsequently diminishes. For higher values of temperature, it can be seen that the measure of entanglement drops completely to zero, in contrast, the measure of quantum correlations decreases but does not vanish. According to these results, we can conclude that, although the quantum entanglement is death, quantum correlations are always present between the two qubits. In the following, Fig. 2 illustrates the dependence of $\mathcal{LN}(\rho)$ and $LQFI$ on the parameter E_{J1} by fixing E_m and E_{J2} at the value of 1.

In Fig. 2, we observe that the quantum indicator $\mathcal{LN}(\rho)$ and $LQFI$ are zero when E_{J1} is zero, regardless of the values of T, meaning that there is no amount of entanglement and no correlations between the two qubits of the system. At $T = 0$, it can be observed that when the value of E_{J1} is lower but not zero, the logarithmic negativity is at its maximum values. Moreover, for a small values of temperature and E_{J1}, the local quantum Fisher information reaches its highest point. An interesting observation is that the two measures decrease with increasing temperature. However, for an important value of T, the two measures do not fail quickly for increased values of E_{J1}. We also notice that $\mathcal{LN}(\rho)$ is more affected by temperature, which disappears for a significant value of temperature, unlike $LQFI$. Hence, $LQFI$ is more resistant against the negative effect of the temperature compared to the logarithmic negativity.

In order to study the influence of E_{J2} on the system of two qubits, we take $E_m = 4$ and $E_{J1} = 10$, with adjusting E_{J2} from 2 to 6. In Fig. 3, the parameter E_{J2} is found to be similar of the effect of E_{J1} on $\mathcal{LN}(\rho)$ and $F(\rho)$.

We see in Fig. 3, that the increase of E_{J2} implies an enhancement in preserving the quantumness of entanglement and correlations between the two superconducting qubits from the influence of the higher values of temperature. Moreover, it should

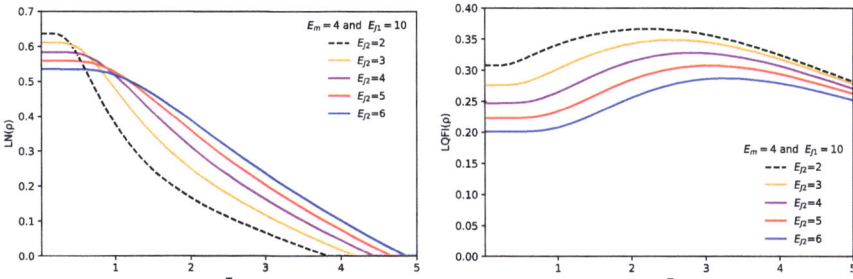

Fig. 3 The behavior of $\mathcal{LN}(\rho)$ and $F(\rho)$ as function of T for various values of E_{J2}, with $E_m = 4$ and $E_{J1} = 10$

be emphasized that $LQFI$ is better than $\mathcal{LN}(\rho)$ at resistant temperature and does not disappear quickly. Consequently, the two measures of quantum resources are impacted by the temperature. Then, we can conclude that $LQFI$ is a good choice to reveal a higher amount of quantum correlations and still long-lived with growing the temperature, unlike $\mathcal{LN}(\rho)$.

References

1. M. Mansour, Z. Dahbi, Entanglement of bipartite partly non-orthogonal-spin coherent states. Laser Phys. **30**, 085201 (2020). https://doi.org/10.1088/1555-6611/ab8c33
2. Z. Dahbi, A.U. Rahman, M. Mansour, Skew information correlations and local quantum Fisher information in two gravitational cat states. Physica A **609**, 128333 (2023). https://doi.org/10.1016/j.physa.2023.128333
3. M.B. Plenio, Logarithmic negativity: a full entanglement monotone that is not convex. Phys. Rev. Lett. **95**, 090503 (2005). https://doi.org/10.1103/PhysRevLett.95.090503
4. Z. Bouafia, M. Mansour, Quantum interferometric power versus quantum correlations in a graphene layer system with a scattering process under thermal noise. Laser Phys. Lett. **20**, 125204 (2023). https://doi.org/10.1088/1612-202X/acf083
5. S. Elghaayda, Z. Dahbi, M. Mansour, Dynamics of quantum correlations in two 2-level atoms coupled to thermal reservoirs. Phys. Scr. **54**, 419 (2022). https://doi.org/10.1088/1402-4896/abc123
6. M. Benzahra, et al. Dynamics of quantum coherence and non-classical correlations between non-interacting two 2-level atoms in thermal baths. Mod. Phys. Lett. A. **39**(16), 2450077 (2024)
7. M.D. Shaw, J.F. Schneiderman, J. Bueno, B.S. Palmer, P. Delsing, P.M. Echternach, Characterization of an entangled system of two superconducting qubits using a multiplexed capacitance measurement. Phys. Rev. B **79**, 014516 (2009). https://doi.org/10.1103/PhysRevB.79.014516
8. M. Benzahra, M. Mansour, M. Oumennana, S. Elghaayda, Quantum correlations and thermal coherence in a two-superconducting charge qubit system. Laser Phys. **33**, 075202 (2023). https://doi.org/10.1088/1555-6611/acf456

Open Access This chapter is licensed under the terms of the Creative Commons Attribution 4.0 International License (http://creativecommons.org/licenses/by/4.0/), which permits use, sharing, adaptation, distribution and reproduction in any medium or format, as long as you give appropriate credit to the original author(s) and the source, provide a link to the Creative Commons license and indicate if changes were made.

The images or other third party material in this chapter are included in the chapter's Creative Commons license, unless indicated otherwise in a credit line to the material. If material is not included in the chapter's Creative Commons license and your intended use is not permitted by statutory regulation or exceeds the permitted use, you will need to obtain permission directly from the copyright holder.

Status and Perspective of ICARUS at the Fermilab Short-Baseline Neutrino Program

Valerio Pia

1 The Short Baseline Neutrino and ICARUS Physics Program

The Short Baseline Neutrino (SBN) program, at Fermilab, aims to either confirm or definitively rule out the potential existence of a light sterile neutrino, suggested by experimental anomalies [1, 2], that participates in the neutrino oscillation without interacting weakly with the three standard neutrinos. SBN is composed of two Liquid Argon Time Projection Chambers (LArTPCs) [3] positioned at different distances along the Booster Neutrino Beam (BNB) and 6° off-axis along the Neutrinos at the Main Injector beam (NuMI) at Fermilab: the Short Baseline Near Detector (SBND) at 110 m, and the Far Detector (FD) ICARUS T600 at 600 m from the neutrino source. Its primary goal is to observe neutrino oscillations in the ν_e appearance and ν_μ disappearance channels at a distance of approximately 1 km and an energy of around 1 GeV.

The combined analysis of data from both the Near and Far detectors will cover the allowed current range in the Δm^2_{14} parameter space with a 5σ sensitivity level in 3 years of data taking [4], corresponding to 6.6×10^{20} POT (Protons On Target, the number of protons delivered by the accelerator to the neutrino generating target), as illustrated in Fig. 1. During the early phases of data taking, ICARUS main goal will be to verify the recent claim of the observation of short-baseline oscillations at $\Delta m^2 \sim 7 \text{ eV}^2$ and $\sin^2_2 \theta \sim 0.36$ reported by the Neutrino-4 reactor experiment

---

Valerio Pia on behalf of the ICARUS Collaboration.

V. Pia (✉)
INFN Sezione di Bologna, Bologna, Italy
e-mail: valerio.pia@bo.infn.it

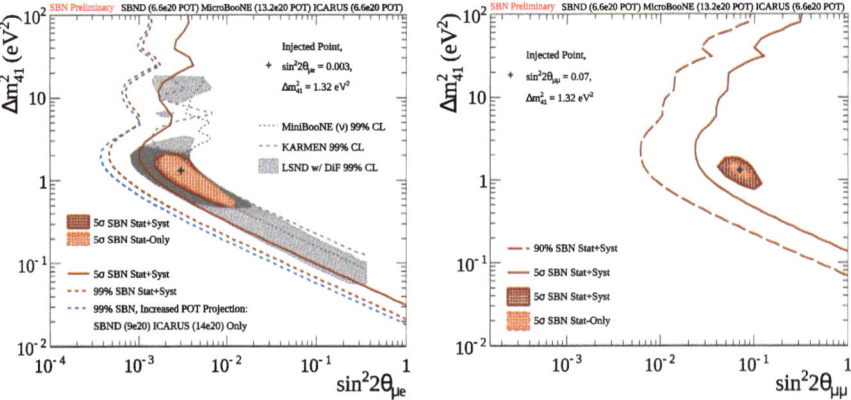

Fig. 1 Expected SBN 5 σ sensitivities (red solid line) to a light sterile neutrino in the ν_e appearance (left) and ν_μ (right) disappearance channel

[5]. ICARUS will study both the ν_e appearance by selecting contained QE ν_e CC candidates, and ν_μ disappearance channels focusing on quasi-elastic ν_μ CC interactions.

2 The ICARUS T600 Detector

ICARUS-T600 consists of two adjacent $3.6 \times 3.9 \times 19.6$ m^3 modules, both filled with a total of 760 tons (476 active tons) of ultra-pure liquid argon (Fig. 2, left). Each module is composed of two Time Projection Chambers (TPC), with a common cathode dividing them, and with a maximum drift distance of 1.5 m. The cathode is set at a voltage of 75 kV, generating a nominal electric field of 500 V/cm, equivalent to $\sim$1 ms drift time. The anode of each TPC is made of three wire planes positioned 3 mm apart and with a different angle with respect to the horizontal direction: 0°, +60°, and −60°. The first two planes perform non-destructive charge measurements, while the last one collects the ionization charge. Additionally, Liquid Argon scintillation light is detected by a light collection system made of 360 Photomultiplier tubes (PMTs) located behind the wire planes, which provide the event time and the trigger of the detector. Due to ICARUS being operated at ground level, an overburden of about 3 m of concrete was installed on top of the detector to reduce the background induced by cosmic rays. Overall, a reduction of a factor 200 was observed for the rate of neutrons and gammas, while a reduction of about 25% was observed for muons. In addition to the overburden, a Cosmic Rays Tagger (CRT) was installed around the detector, providing a 4π coverage of the detector with 95% tagging efficiency and few ns time resolution (Fig. 2, right). The CRT system comprises modules of plastic scintillator bars with embedded Wavelength Shifter (WLS) fibers coupled to Silicon Photomultipliers (SiPM).

Fig. 2 Left: Deployment of the second ICARUS module inside the pit of the SBN Far Detector experimental hall at Fermilab. Right: CRT modules installed on top of the detector. *Reprinted under CC-BY-4.0 license from [6]. ©2023, ICARUS Collaboration*

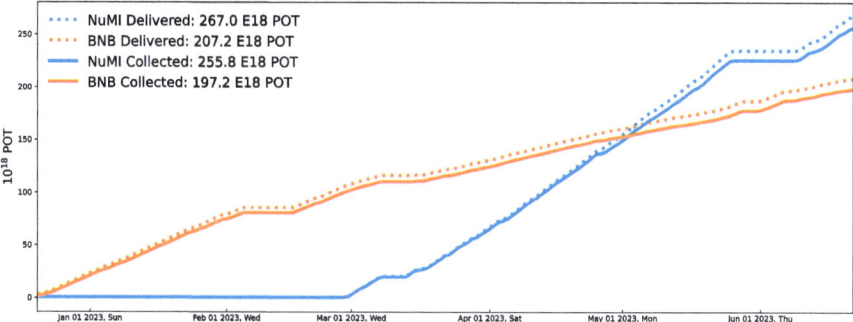

Fig. 3 Cumulative sum of POT delivered by the accelerator (dotted lines) and collected by the detector (solid lines) as a function of the operation time for BNB (orange lines) and NuMI (blue lines) for the ICARUS Run 2, started December 2022 and completed June 2023

3 ICARUS Operations

The ICARUS detector was fully operational in June 2021 and started a commissioning phase that lasted until June 2022 [6]. During the commissioning, it collected 296×10^{18} POT and 503×10^{18} POT collected for BNB and NuMI, respectively with a beam utilization—defined as the amount of POT collected divided by the amount delivered—of 89% for BNB and 88% for NuMI. In June 2022, ICARUS started taking physics quality data, with a first run lasting until July 2022 (Run 1), and a second one (Run 2) started in December 2022 and lasted until July 2023, just before the summer beam downtime. The combined statistic of both runs is of 2.46×10^{20} POT for BNB and 3.42×10^{20} POT for NuMI, with a beam utilization of 93% and 95% respectively. Figure 3 shows the cumulative sum of POT delivered by the accelerator and collected by the detector for the full Run 2 period and both beams.

4 Observation and Reconstruction of Neutrino Events

The data collected by the detector is processed to reconstruct and analyze the events. The signal obtained from the wires is reconstructed by performing a deconvolution of the waveforms to obtain the structure of the drift electrons generating the signal at the moment of their production. The ICARUS deconvolution procedure correctly associates a wire signal with the track generating it with an efficiency higher than 90% for track segments with lengths larger than 3.4 mm [6]. The scintillation light associated with an event is reconstructed from the signal recorded by the PMTs. A threshold-based algorithm is applied to each recorded signal, to reconstruct the characteristics of the detected light to be used in the analysis. Initially, the association of an event in the TPC with the light detected by the PMTs was performed comparing the track and the light barycentre along the longitudinal z-axis. A correlation within a few tens of centimeters was observed for both cosmic muons (Fig. 4, left) and for a sample of BNB neutrino interactions selected by visual scanning (Fig. 4, right) [6]. CRT hits are reconstructed by searching for coincidences of signals above a threshold on different planes of the CRT modules. For each found coincidence, spatial and timing information of the hit are provided.

Reconstructed hits in each cryostat are then passed as input to a pattern reconstruction code, Pandora [7], that identifies interaction vertices, and reconstructs tracks and showers inside the TPC by performing a full 3D reconstruction of the event. From initial Monte Carlo studies, a rejection of 99.7% of cosmic rays was obtained, while also accepting more than 82% of true ν_μCC events. Pandora, together with new algorithms developed jointly by SBN, are applied to tracks and showers reconstructed in the event to perform particle identification and reconstruction, including the reconstruction of the momentum, calorimetry, and contribution of multiple Coulomb scattering. To validate the Padora reconstruction, a dedicated visual study of ∼500 ν_μ CC interaction from BNB was performed. In ∼90% of these events, the difference of the neutrino interaction vertex along the beam direction reconstructed manually and by Pandora is within 3 cm. Comparison between visual studies and automated reconstructions will help improve the auto-

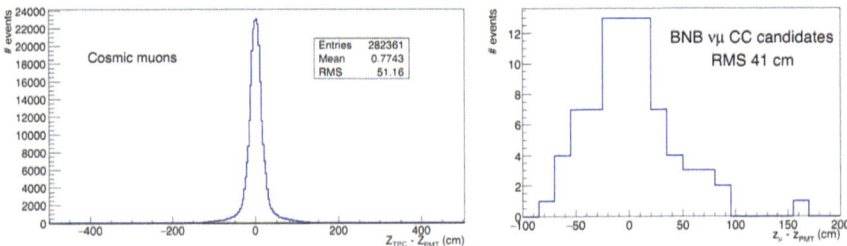

Fig. 4 Left: distribution of $\Delta z = z_{TPC} - z_{PMT}$ for a sample of cosmic ray muons crossing the cathode. Right: same distribution for a sample of BNB ν interactions identified by visual scanning. *Reprinted under CC-BY-4.0 license from [6]. ©2023, ICARUS Collaboration*

matic reconstruction. However, the first results are quite promising and demonstrate that the basic tools for event selection and reconstruction are operational and allow an initial identification and measurement of neutrino interactions.

5 Conclusions

The ICARUS detector has been fully operational since June 2021 and collected a total of 2.46×10^{20} POT and 3.42×10^{20} POT from the BNB and NuMI beams respectively. Automated selection and reconstructed algorithms are currently being tuned and will allow sensitive studies of the claim by the Neutrino-4 short-baseline reactor experiment. Within the Short-Baseline Neutrino program, ICARUS will then conduct searches for evidence of a sterile neutrino.

References

1. A. Aguilar et al., Evidence for neutrino oscillations from the observation of $\bar{\nu}_e$ appearance in a $\bar{\nu}_\mu$ beam. Phys. Rev. D **64**(11) (2001). https://doi.org/10.1103/PhysRevD.64.112007
2. G. Mention et al., Reactor antineutrino anomaly. Phys. Rev. D **83**(7) (2011). https://doi.org/10.1103/PhysRevD.83.073006
3. C. Rubbia, The liquid-argon time projection chamber: a new concept for neutrino detectors (1977). Preprint at https://cds.cern.ch/record/117852
4. P.A.N. Machado, O. Palamara, D.W. Schmitz, The short-baseline neutrino program at Fermilab. Annu. Rev. Nucl. Particle Sci. **69**, 363–387 (2019). https://doi.org/10.1146/annurev-nucl-101917-020949
5. A.P. Serebrov et al., Search for sterile neutrinos with the Neutrino-4 experiment and measurement results. Phys. Rev. D **104**(3) (2021). https://doi.org/10.1103/PhysRevD.104.032003
6. P. Abratenko et al., ICARUS at the Fermilab Short-Baseline Neutrino program: initial operation. Eur. Phys. J. C **83**, 467 (2023). https://doi.org/10.1140/epjc/s10052-023-11610-y
7. R. Acciarri et al., The Pandora multi-algorithm approach to automated pattern recognition of cosmic-ray muon and neutrino events in the MicroBooNE detector. Eur. Phys. J. C **78**, 82 (2018). https://doi.org/10.1140/epjc/s10052-017-5481-6

Open Access This chapter is licensed under the terms of the Creative Commons Attribution 4.0 International License (http://creativecommons.org/licenses/by/4.0/), which permits use, sharing, adaptation, distribution and reproduction in any medium or format, as long as you give appropriate credit to the original author(s) and the source, provide a link to the Creative Commons license and indicate if changes were made.

The images or other third party material in this chapter are included in the chapter's Creative Commons license, unless indicated otherwise in a credit line to the material. If material is not included in the chapter's Creative Commons license and your intended use is not permitted by statutory regulation or exceeds the permitted use, you will need to obtain permission directly from the copyright holder.

Revisiting Inert Doublet Model Parameters

Hamza Abouabid, Abdesslam Arhrib, Ayoub Hmissou, and Larbi Rahili

1 Introduction

After the discovery of the Higgs particle around 125 GeV, a wide window opened in the field of particle physics for exploring the properties of the Higgs boson. This discovery demonstrated that the SM effectively explains observed phenomena at the electroweak scale. However, the SM fails to address several issues, such as non-zero neutrino masses, dark matter (DM), and gravitational interactions. This has spurred the search for new physics beyond the Standard Model (BSM), often by extending the SM to include additional real or complex singlets, doublets, or triplet Higgs fields.

Recently, an excess has been observed for the di-photon Higgs decay, which we are attempting to explain within the framework of the Inert Doublet model. In this proceeding examines how current Di-photon ($\mu_{\gamma\gamma}$) and recent Z-photon ($\mu_{\gamma Z}$) signal measurements constrain the IDM parameter space, particularly the coupling hH^+H^-. Additionally, it provides updated insights into the impact of extra scalars on radiative corrections to the triple Higgs coupling hhh at the one-loop level.

H. Abouabid · A. Arhrib
Abdelmalek Essaadi University, Faculty of Sciences and Techniques, Tangier, Morocco

A. Hmissou (✉) · L. Rahili
Laboratory of Theoretical and High Energy Physics, Faculty of Science, Ibnou Zohr University, Agadir, Morocco
e-mail: ayoub.hmissou@edu.uiz.ac.ma

2 The Model

The Inert Doublet Model (IDM) extends the Standard Model (SM) Higgs sector by introducing a dark matter (DM) candidate [1–3]. It is a variant of the 2HDM with an exact Z_2 symmetry. This involves adding an inert scalar doublet, H_2, to the SM Higgs doublet, H_1. Notably, H_2 has odd parity under the Z_2 symmetry, resulting in no coupling with fermions and no vacuum expectation value. The physical parametrization of the scalar doublets has the form

$$H_1 = \begin{pmatrix} G^+ \\ (v+h+iG)/\sqrt{2} \end{pmatrix}, \quad H_2 = \begin{pmatrix} H^+ \\ (S+iA)/\sqrt{2} \end{pmatrix}, \quad (1)$$

The Higgs potential is given by:

$$V_{\text{IDM}} = \mu_{11}^2 H_1^\dagger H_1 + \mu_{22}^2 H_2^\dagger H_2 + \eta_1 \left(H_1^\dagger H_1\right)^2 + \eta_2 \left(H_2^\dagger H_2\right)^2$$
$$+ \eta_3 \left(H_1^\dagger H_1\right)\left(H_2^\dagger H_2\right)$$
$$+ \eta_4 \left(H_1^\dagger H_2\right)\left(H_2^\dagger H_1\right) + \frac{1}{2}\eta_5 \left[\left(H_1^\dagger H_2\right)^2 + \left(H_2^\dagger H_1\right)^2\right], \quad (2)$$

Where μ_{11}^2 and μ_{22}^2 represent mass terms, and $\eta_{i=1...5}$ denote quartic couplings. When the electroweak symmetry is broken, H_1 acquires its vacuum expectation value VEV $<H_1>_0 = v$ GeV, while H_2 maintains a vanishing VEV $<H_2>_0 = 0$. The IDM consists of six free parameters that we have chosen as:

$$\mathcal{P} = \{\mu_{22}^2, \eta_2, m_h, m_S, m_A, m_{H^\pm}\} \quad (3)$$

The scalar bosons masses are given by:

$$m_h^2 = 2\eta_1 v^2 = -2\mu_{11}^2, \quad (4)$$
$$m_S^2 = \mu_{22}^2 + \eta_L v^2, \quad (5)$$
$$m_A^2 = \mu_{22}^2 + \eta_S v^2, \quad (6)$$
$$m_{H^\pm}^2 = \mu_{22}^2 + \frac{1}{2}\eta_3 v^2. \quad (7)$$

where the new expressions $\eta_{L,S}$ are as follows

$$\eta_L = \frac{1}{2}(\eta_3 + \eta_4 + \eta_5), \quad \eta_S = \frac{1}{2}(\eta_3 + \eta_4 - \eta_5) \quad (8)$$

3 Results and Discussions

This section explores the impact of new measurement on the parameter space and the trilinear coupling η_{hhh} within the IDM. They applied renormalisation techniques to following process

$$h(q) \to h(k_1) + h(k_2) \tag{9}$$

and showed its sensitivity to the effects of NP BSM. Here, with the aim of approaching IDM towards such measures, we revisit the additional contribution of the following process.

In line with our purpose in this study, we redefine a ratio that involves the previous quantities as,

$$\Delta \Gamma_{hhh} = \frac{\Gamma_{hhh}^{IDM,1} - \Gamma_{hhh}^{IDM,0}}{\Gamma_{hhh}^{IDM,0}} \tag{10}$$

where $\Gamma_{hhh}^{IDM,1}$ is the one loop amplitude and $\Gamma_{hhh}^{IDM,0} = -3m_h^2/v$ represents the trilinear coupling at tree level, identical to the SM value. For more details, we refer the reader to [4, 5].

In the numerical scan, we will take into consideration all theoretical constraints Perturbativity, Vacuum Stability, Charge-breaking minima, Inert Vacuum and Unitarity on the scalar sector of the model as well as LHC experimental measurement such $\mu_{\gamma\gamma}^{exp}$ and $\mu_{Z\gamma}^{exp}$ and the invisible decay of the SM Higgs, as detailed in Ref. [5]. Also, all along our computings, latest version of HiggsBounds-5.10.2. [6] and HiggsSignals-2.6.2. [7] and micromegas-5. [8] have been used.

Figure 1 displays permissible parameter space, indicating viable ranges for m_Φ and η_3. At 99%C.L., m_Φ is constrained to $\leq$500 GeV and $\eta_3 \leq 9$, confirmed by

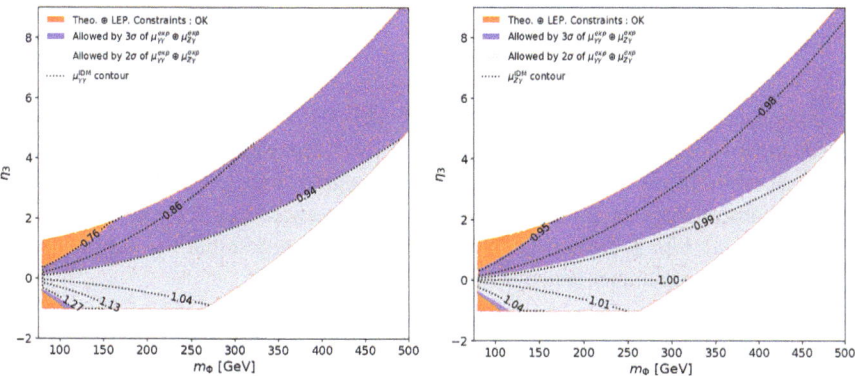

Fig. 1 The correlation between $m_\phi = m_S = m_A = m_{H^\pm}$ and η_3 with the dot-dashed line corresponds to the $\mu_{\gamma\gamma}^{\text{IDM}}$ (left) and $\mu_{Z\gamma}^{\text{IDM}}$ (right) contours

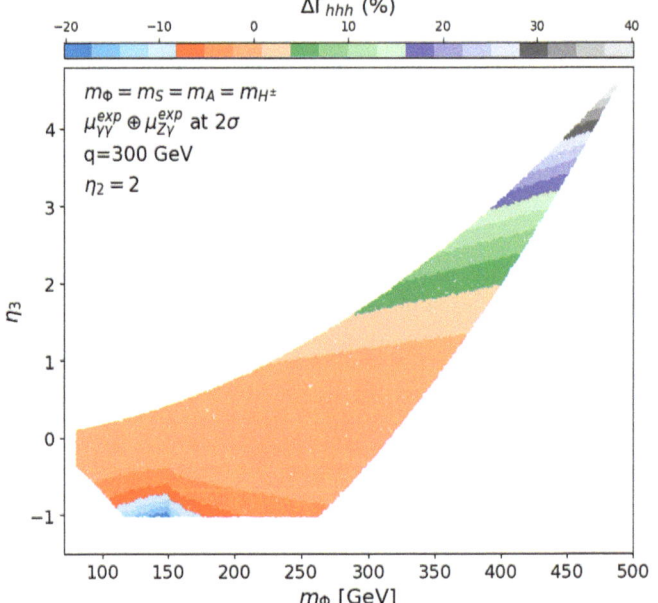

Fig. 2 The $\Delta\Gamma_{hhh}$ variation within the $(\eta_3 - m_\Phi)$ plane at 95% .C.L. with $\eta_2 = 2$

the gray region. Narrowing to $\pm 2\sigma$ reveals a smaller area (orange) up to 488 GeV, ruling out enhancements for $Br(h \to Z\gamma)/SM$ and $Br(h \to \gamma\gamma)/SM$ above 4.5% and 27%, respectively. Experimental validation requires a positive μ_{22}^2.

Figure 2 focuses on $\mu_{\gamma\gamma,\gamma Z}^{exp}$ at 2σ C.L., showing reduced parameter space and diminished radiative corrections, particularly for m_Φ near 485 GeV. $|\Delta\Gamma_{hhh}|$ increases steadily over η_3, peaking around $\eta_3 = -1$ for $m_\Phi \approx 150$ GeV.

4 Conclusion

In this paper, we revisit the Inert Doublet Model (IDM) as an extension of Beyond the Standard Model (BSM), focusing on the two decay modes, $\gamma\gamma$ and $Z\gamma$. We examine the implications of current data on the signal strengths $\mu_{\gamma\gamma,\gamma Z}^{exp}$ for various physical observables. in addition the effects on the radiative corrections to the triple Higgs coupling hhh at the one-loop level. Our results suggest that $\gamma\gamma$ may constrain the permissible parameter space for the IDM.

References

1. N.G. Deshpande, E. Ma, Pattern of symmetry breaking with two Higgs doublets. Phys. Rev. D **18**, 2574 (1978). https://doi.org/10.1103/PhysRevD.18.2574
2. Q.H. Cao, E. Ma, G. Rajasekaran, Observing the dark scalar doublet and its impact on the standard-model Higgs Boson at colliders. Phys. Rev. D **76**, 095011 (2007). https://doi.org/10.1103/PhysRevD.76.095011
3. R. Barbieri, L.J. Hall, V.S. Rychkov, Improved naturalness with a heavy Higgs: an alternative road to LHC physics. Phys. Rev. D **74**, 015007 (2006). https://doi.org/10.1103/PhysRevD.74.015007
4. A. Arhrib, R. Benbrik, J. El Falaki, A. Jueid, Radiative corrections to the triple Higgs coupling in the inert Higgs doublet model. J. High Energy Phys. **12**, 007 (2015). https://doi.org/10.1007/JHEP12(2015)007
5. H. Abouabid, A. Arhrib, A. Hmissou, L. Rahili, Revisiting inert doublet model parameters. Eur. Phys. J. C **84**(6), 632 (2024). https://doi.org/10.1140/epjc/s10052-024-13014-y
6. P. Bechtle, O. Brein, S. Heinemeyer, O. Stål, T. Stefaniak, G. Weiglein, K.E. Williams, HiggsBounds − 4: improved tests of extended Higgs sectors against exclusion bounds from LEP, the Tevatron and the LHC. Eur. Phys. J. C **74**(3), 2693 (2014). https://doi.org/10.1140/epjc/s10052-013-2693-2
7. P. Bechtle, S. Heinemeyer, O. Stål, T. Stefaniak, G. Weiglein, *HiggsSignals*: confronting arbitrary Higgs sectors with measurements at the Tevatron and the LHC. Eur. Phys. J. C **74**(2), 2711 (2014). https://doi.org/10.1140/epjc/s10052-013-2711-4
8. G. Belanger, F. Boudjema, A. Pukhov, A. Semenov, micrOMEGAs: a tool for dark matter studies. Nuovo Cim. C **033N2**, 111–116 (2010). https://doi.org/10.1393/ncc/i2010-10591-3

Open Access This chapter is licensed under the terms of the Creative Commons Attribution 4.0 International License (http://creativecommons.org/licenses/by/4.0/), which permits use, sharing, adaptation, distribution and reproduction in any medium or format, as long as you give appropriate credit to the original author(s) and the source, provide a link to the Creative Commons license and indicate if changes were made.

The images or other third party material in this chapter are included in the chapter's Creative Commons license, unless indicated otherwise in a credit line to the material. If material is not included in the chapter's Creative Commons license and your intended use is not permitted by statutory regulation or exceeds the permitted use, you will need to obtain permission directly from the copyright holder.

Conformity of New CDF-II M_W with 123-Model

B. Ait Ouazghour, R. Benbrik ⓘ, Es-said Ghourmin, M. Ouchemhou, and L. Rahili

1 Introduction

High-precision measurements at collider experiments have imposed stringent constraints on the Standard Model (SM) and its possible extensions. This level of precision has been further reinforced by the discovery of the Higgs boson at the LHC experiment [1, 2]. Recently, the CDF collaboration reported a newly measured value of the W boson mass [3], which deviates from the SM prediction by approximately 7σ. The SM expectation, as provided by [4, 5], is given by:

$$m_W^{\text{CDF}} = 80.4335 \pm 0.0094 \text{ GeV}, \quad m_W^{\text{SM}} = 80.357 \pm 0.006 \text{ GeV}. \tag{1}$$

In this work, we focus on the 123-model to investigate the possibility of predicting m_W according to the new CDF measurement.

The correction of the 123-model to the W boson mass and the effective weak mixing angle can be parameterized using the S, T, and U formalism as follows [6]:

$$\Delta m_W^2 = \left(m_W^{123}\right)^2 - \left(m_W^{\text{SM}}\right)^2 = \frac{\alpha_0 c_W^2 m_Z^2}{c_W^2 - s_W^2} \left[-\frac{1}{2} S + c_W^2 T + \frac{c_w^2 - s_w^2}{4 s_W^2} U \right], \tag{2}$$

B. Ait Ouazghour
LPHEA, Faculty of Science Semlalia, Cadi Ayyad University, Marrakech, Morocco

R. Benbrik · M. Ouchemhou
Polydisciplinary Faculty, Laboratory of Fundamental and Applied Physics, Cadi Ayyad University, Sidi Bouzid, Safi, Morocco

E. Ghourmin (✉) · L. Rahili
Laboratory of Theoretical and High Energy Physics, Faculty of Science, Ibn Zohr University, Agadir, Morocco

$$\Delta \sin^2 \theta_{\text{eff}} = \sin^2 \theta_{\text{eff}}\Big|_{123} - \sin^2 \theta\Big|_{\text{SM}} = \frac{\alpha_0}{c_W^2 - s_W^2}\left[\frac{1}{4}S - s_W^2 c_W^2 T\right], \quad (3)$$

where the SM values used are listed in Ref. [4].

This work is organised as follows. In Sect. 2, we describe in great length the 123 model. Our main results are discussed in Sect. 3 and finally, we summarize our conclusions in Sect. 4.

2 123-Model in a Nutshell

In addition to the usual SM scalar doublet, namely ϕ, a singlet σ, and a triplet Δ have been introduced as fundamental building blocks of the 123-model [10]. Considering their representations under the $SU(3)c \times SU(2)L \times U(1)_Y$ SM gauge group, one can explicitly express them as follows:

$$\sigma(1, 1, +0) = \frac{1}{\sqrt{2}}(v_\sigma + R_\sigma + iI_\sigma),$$

$$\phi(1, 2, +\frac{1}{2}) = \begin{pmatrix} \frac{1}{\sqrt{2}}(v_\phi + R_\phi + iI_\phi) \\ \phi^- \end{pmatrix},$$

$$\Delta(1, 3, +1) = \begin{pmatrix} \frac{1}{\sqrt{2}}(v_\Delta + R_\Delta + iI_\Delta) & \Delta^+/\sqrt{2} \\ \Delta^+/\sqrt{2} & \Delta^{++} \end{pmatrix}, \quad (4)$$

with corresponding leptonic numbers $L_\sigma = 2$, $L_\phi = 0$ and $L_\Delta = -2$, respectively.

The most general renormalizable and gauge-invariant Lagrangian of the 123 scalar sector is given by

$$\mathcal{L} = (D_\mu \phi)^\dagger (D^\mu \phi) + Tr(D_\mu \Delta)^\dagger (D^\mu \Delta) + (\partial_\mu \sigma)^\dagger (\partial^\mu \sigma) - V(H, \Delta) + \mathcal{L}_{\text{Yukawa}}, \quad (5)$$

with the covariant derivatives in the kinetic terms are

$$D_\mu \phi = \partial_\mu \phi + igT^a W_\mu^a \phi + i\frac{1}{2}g' B_\mu \phi,$$

$$D_\mu \Delta = \partial_\mu \Delta + ig[T^a W_\mu^a, \Delta] + ig'\frac{Y_\Delta}{2} B_\mu \Delta, \quad (6)$$

where W_μ^a and B_μ stand for $SU(2)_L$ and $U(1)_Y$ gauge fields, respectively. Y_Δ is the hypercharge operator of the triplet Δ, while T^a is related to the Pauli matrices via $T^a = \sigma^a/2$. $\mathcal{L}_{\text{Yukawa}}$ refer to the Yukawa part to be subsequently considered in detail.

The scalar potential $V(\sigma, \phi, \Delta)$, invariant under $SU(2)_L \times U(1)_Y$, reads [7]

$$V(\sigma, \phi, \Delta) = \mu_\sigma^2 \sigma^\dagger \sigma + \mu_\phi^2 \phi^\dagger \phi + \mu_\Delta^2 \text{Tr}(\Delta^\dagger \Delta) + \lambda_1 (\phi^\dagger \phi)^2$$
$$+ \lambda_2 \left[\text{Tr}(\Delta^\dagger \Delta)\right]^2 + \lambda_3 (\phi^\dagger \phi) \text{Tr}(\Delta^\dagger \Delta)$$
$$+ \lambda_4 \text{Tr}(\Delta^\dagger \Delta \Delta^\dagger \Delta) + \lambda_5 (\phi^\dagger \Delta^\dagger \Delta \phi) + \beta_1 (\sigma^\dagger \sigma)^2$$
$$+ \beta_2 (\phi^\dagger \phi)(\sigma^\dagger \sigma) + \beta_3 \text{Tr}(\Delta^\dagger \Delta)(\sigma^\dagger \sigma)$$
$$- \kappa (\phi^T \Delta \phi \sigma + h.c.), \tag{7}$$

where all quartic couplings are considered to be real. μ_i^2 ($i = \sigma, \phi, \Delta$) are squared mass parameters of the singlet, doublet, and triplet fields, respectively.

3 Result and Discussion

To study the implication of the new CDF measurement on the 123-Model, we perform a systematic scan over its parameter space. The scan is done using the our model code. We assume that the CP-even Higgs boson h_1 is the observed SM-like Higgs with M_{h_1} = 125.09 GeV whose properties are consistent with the LHC measurements. We randomly sample the remaining model parameters within the ranges given in Table 1.

During the scan, the following theoretical and experimental constraints are fulfilled:

- Unitarity, vacuum stability, and perturbativity requirements.
- The compatibility with the 95% bounds imposed by the LHC is checked via the public program HiggsBounds-5.10.2 [8].
- The criterion that the CP-even h_1 Higgs boson must match the characteristics of the observed SM-like Higgs boson is enforced using the public code HiggsSignal-2.6.2. [9].
- Electroweak precision observables (EWPO) are examined using the oblique parameters S and T (with U fixed to zero), based on both the PDG [4] and CDF [3] fit results. In particular, we perform the χ^2_{ST} test prior to and following the new m_W^{CDF} measurement, referred to as "PDG" and "CDF," respectively.

$$\text{PDG}: S = 0.05 \pm 0.08, \ T = 0.09 \pm 0.07, \ \rho_{ST} = 0.92 \tag{8}$$
$$\text{CDF}: S = 0.15 \pm 0.08, \ T = 0.27 \pm 0.06, \ \rho_{ST} = 0.93, \tag{9}$$

where ρ_{ST} represents the correlation between S and T.

Once the allowed parameter space, which satisfies the outlined theoretical and experimental constraints, is determined, we proceed to conduct the $\chi^2_{M_W^{\text{CDF}}}$ test. This analysis focuses exclusively on points that fall within 2σ of the new CDF measurement.

Table 1 123-Parameter space scan where h_1 is SM-like (all *vev*'s are in GeV). Reprinted under CC-BY-4.0 from [10]. ©2023, The Author(s)

λ_2	λ_4	λ_5	β_3	κ	v_Δ	v_s	α_1	α_2	α_3
$[-6;8\pi]$	$[-9;8\pi]$	$[-15;14]$	$[-8\pi;8\pi]$	$[0;0.1]$	$[0;1]$	$[10;1000]$	$[-\pi/2;\pi/2]$	$[-0.5;0.5]$	$[-\pi/2;\pi/2]$

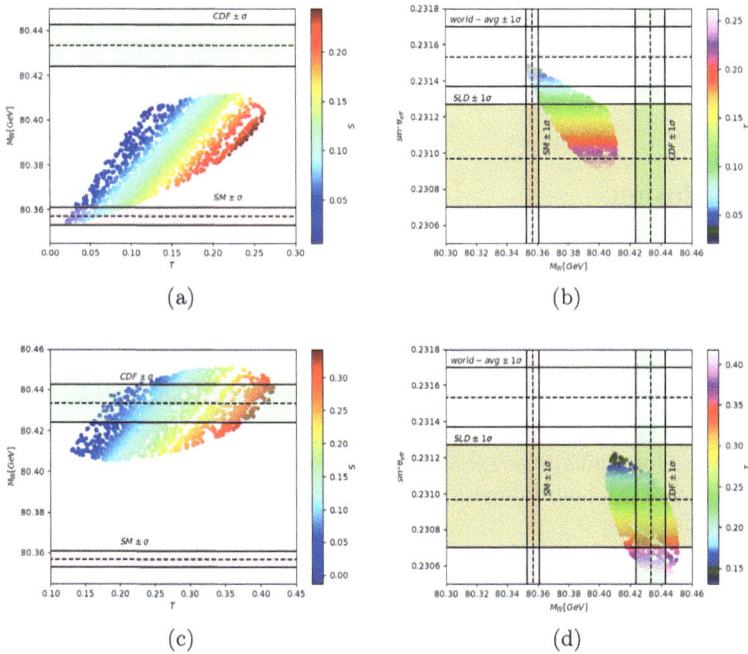

Fig. 1 Left panels: The estimation of m_W in the 123-model as a function of T, with the color gradient representing the magnitude of the parameter S. The light pink bands correspond to the SM prediction, while the green bands indicate the new CDF measurement of m_W within a 1σ uncertainty. Right panels: The prediction of $\sin^2\theta_{\text{eff}}$ in the 123-model as a function of m_W, where the color gradient represents the magnitude of T. The light blue band denotes the world average value of $\sin^2\theta_{\text{eff}}$ with its corresponding 1σ uncertainty, while the brown region shows the SLD collaboration result at the 1σ level. The upper (lower) panels correspond to the PDG (CDF) fit results. Reprinted under CC-BY-4.0 from [10]. ©2023, The Author(s)

The preliminary summary results are presented in Fig. 1. Considering the PDG values, Fig. 1a displays the 123-prediction for m_W as a function of T, with the color mapping representing the S parameter. At first glance, it is evident that the predicted m_W value within the 123-model is consistent with the SM prediction, requiring $T \in [0.02, 0.10]$ and $S \in [0, 0.15]$, while showing a significant deviation from the new CDF region at the 1σ level.

However, when χ^2_{STU}(PDG) is replaced with χ^2_{STU}(CDF) in the global χ^2, the CDF bands become interpretable within the 123-model. This adjustment allows the Peskin parameters T and S to marginally fall within the ranges 0.15–0.42 and 0–0.35, respectively, as shown in Fig. 1c.

On the other hand, the right panel in Fig. 1 presents the model prediction for m_W as a function of $\sin^2\theta_{\text{eff}}$. As depicted in Fig. 1d, when considering the CDF S, T results, the parameter points predominantly align with the SLD measurement for $\sin^2\theta_{\text{eff}}$ within the 1–1.5σ range. This contrasts with the scenario using the PDG values, where the predicted value exhibits consistency with both experimental measurements.

4 Conclusion

The CDF II experiment has reported a significant anomaly in the measured mass of the W boson. With remarkable precision, the result reveals a slightly higher value than the Standard Model (SM) prediction. This intriguing deviation is currently under active investigation, as it may provide evidence for new physics beyond the Standard Model (BSM). In this paper, we have explored the compatibility of this anomaly within the 123-model, considering both theoretical and experimental constraints. Our findings suggest that the newly measured m_W at the CDF-II experiment favors a nonzero mass splitting among the particles h_2, h_3, A, $H^\pm$, and $H^{\pm\pm}$ [10].

References

1. ATLAS Collaboration, G. Aad et al., Observation of a new particle in the search for the Standard Model Higgs boson with the ATLAS detector at the LHC. Phys. Lett. B **716**, 1–29 (2012) [arXiv:1207.7214]
2. CMS Collaboration, S. Chatrchyan et al., Observation of a new Boson at a mass of 125 GeV with the CMS experiment at the LHC. Phys. Lett. B **716**, 30–61 (2012) [arXiv:1207.7235]
3. CDF Collaboration, T. Aaltonen et al., High-precision measurement of the W boson mass with the CDF II detector. Science **376**(6589), 170–176 (2022)
4. Particle Data Group Collaboration, P.A. Zyla et al., Review of particle physics. Prog. Theor. Exp. Phys. **2020**(8), 083C01 (2020)
5. M. Awramik, M. Czakon, A. Freitas, G. Weiglein, Precise prediction for the W boson mass in the standard model. Phys. Rev. D **69**, 053006 (2004) [hep-ph/0311148]
6. M.E. Peskin, T. Takeuchi, Estimation of oblique electroweak corrections. Phys. Rev. D **46**, 381–409 (1992)
7. A.G. Akeroyd, M.A. Diaz, M.A. Rivera, D. Romero, Fermiophobia in a Higgs triplet model. Phys. Rev. D **83**, 095003 (2011) [arXiv:1010.1160]
8. P. Bechtle, D. Dercks, S. Heinemeyer, T. Klingl, T. Stefaniak, G. Weiglein, J. Wittbrodt, HiggsBounds-5: testing Higgs sectors in the LHC 13 TeV Era. Eur. Phys. J. C **80**(12), 1211 (2020) [arXiv:2006.06007]
9. P. Bechtle, S. Heinemeyer, T. Klingl, T. Stefaniak, G. Weiglein, J. Wittbrodt, HiggsSignals-2: probing new physics with precision Higgs measurements in the LHC 13 TeV era. Eur. Phys. J. C **81**(2), 145 (2021) [arXiv:2012.09197]
10. B. Ait Ouazghour, R. Benbrik, E. Ghourmin, M. Ouchemhou, L. Rahili, Consistency of new CDF-II W boson mass with 123-model. Phys. Lett. B (2023) [arXiv:2302.01641]

Open Access This chapter is licensed under the terms of the Creative Commons Attribution 4.0 International License (http://creativecommons.org/licenses/by/4.0/), which permits use, sharing, adaptation, distribution and reproduction in any medium or format, as long as you give appropriate credit to the original author(s) and the source, provide a link to the Creative Commons license and indicate if changes were made.

The images or other third party material in this chapter are included in the chapter's Creative Commons license, unless indicated otherwise in a credit line to the material. If material is not included in the chapter's Creative Commons license and your intended use is not permitted by statutory regulation or exceeds the permitted use, you will need to obtain permission directly from the copyright holder.

Enhanced Photon Higgs Associated Production in Two Higgs Doublet Type-II Seesaw Model at e^-e^+ Colliders

B. Ait Ouazghour, M. Chabab, and Khalid Goure

1 Introduction

The discovery of the Higgs boson by the ATLAS and CMS collaborations, the last missing piece in the completion of the SM, provided an experimental evidence for the Brout-Englert-Higgs mechanism. Although most of the SM's predictions have undergone successful testing with a high degree of accuracy [1–3] it failed to explain several established physical phenomena. Moreover, given the absence of any direct evidence supporting new physics so far, precision measurements of the Higgs boson properties and couplings to other new scalars can offer a promising opportunity for a potential discovery of new physics. The primary objective of future e^-e^+ colliders [4], such as the International Linear Collider (ILC) [5, 6], Compact Linear Collider (CLIC) [4, 7], Circular Electron-Positron Collider (CEPC) [8–10], and Future Circular Collider (FCC) [11, 12] is to enhance our comprehension of the Higgs physics and refine measurements of the Higgs boson couplings [13–17] with exceptional precision. Compared to hadron colliders, these colliders, featuring a cleaner e^+e^- background, can yield substantial improvements over LHC measurements . The associated production of SM-like Higgs boson with a photon, $e^+e^- \to h_1\gamma$, is well suited to study the Higgs-gauge bosons couplings such as the $h_1\gamma\gamma$ and $h_1\gamma Z$ couplings. This process was investigated in the SM [18–20] and recently in some beyond the Standard Model (BSM) scenarios as the inert Higgs doublet model [21], Higgs Triplet Model (HTM) [22], the minimal supersymmetric standard model (MSSM) [23, 24] and the effective field theory [25].

In this contribution, we investigate the single production of the neutral Higgs boson in association with a photon in electron positron collisions in the frame-

---

B. Ait Ouazghour · M. Chabab · K. Goure (✉)
LPHEA, Faculty of Science Semlalia, Cadi Ayyad University, Marrakech, Morocco
e-mail: brahim.aitouazghour@edu.uca.ac.ma; mchabab@uca.ac.ma; k.goure.ced@uca.ac.ma

work of 2HDMcT (For details see [26]). To do that, we have implemented a full set of theoretical constraints originated from `perturbative unitarity`, `electroweak vacuum stability`. Given that any BSM scenario must be capable of accommodating a 125 GeV state h_{125}, that exhibits properties consistent with present measurements, to insure that compatibility the `HiggsTools` package [27] is employed. This guarantees that the allowed parameter space align with the observed properties of the 125 GeV Higgs boson (`HiggsSignals` [27–30]) and with the limits from searches for additional Higgs bosons at the LHC and at LEP (`HiggsBounds` [27, 31–34]). In addition the parameter space of the model is further constrained using `Electroweak precision observables` and the $\bar{B} \to X_s \gamma$ constraint at 95% C.L.

The structure of this contribution is as follows : In Sect. 2, we briefly introduce the 2HDMcT. Section 3 presents the various theoretical and experimental constraints imposed on the model parameter space, followed by a discussion of our results. Finally, Sect. 4 is dedicated to our conclusion.

2 General 2HDMcT

The most general $SU(2)_L \times U(1)_Y$ invariant scalar potential for 2HDMcT can be written as [35]:

$$V(\Phi_1, \Phi_2, \Delta) = m_{11}^2 \Phi_1^\dagger \Phi_1 + m_{22}^2 \Phi_2^\dagger \Phi_2 - [m_{12}^2 \Phi_1^\dagger \Phi_2 + \text{h.c.}]$$
$$+ \frac{\lambda_1}{2}(\Phi_1^\dagger \Phi_1)^2 + \frac{\lambda_2}{2}(\Phi_2^\dagger \Phi_2)^2$$
$$+ \lambda_4 (\Phi_1^\dagger \Phi_2)(\Phi_2^\dagger \Phi_1)$$
$$+ \left\{ \frac{\lambda_5}{2}(\Phi_1^\dagger \Phi_2)^2 + \left[\beta_1(\Phi_1^\dagger \Phi_1) + \beta_2(\Phi_2^\dagger \Phi_2) \right] \Phi_1^\dagger \Phi_2 + \text{h.c.} \right\}$$
$$+ \lambda_3(\Phi_1^\dagger \Phi_1)(\Phi_2^\dagger \Phi_2) + \lambda_6 \Phi_1^\dagger \Phi_1 Tr \Delta^\dagger \Delta + \lambda_7 \Phi_2^\dagger \Phi_2 Tr \Delta^\dagger \Delta$$
$$+ \left\{ \mu_1 \Phi_1^T i\sigma^2 \Delta^\dagger \Phi_1 + \mu_2 \Phi_2^T i\sigma^2 \Delta^\dagger \Phi_2 + \mu_3 \Phi_1^T i\sigma^2 \Delta^\dagger \Phi_2 + \text{h.c.} \right\}$$
$$+ \lambda_8 \Phi_1^\dagger \Delta \Delta^\dagger \Phi_1$$
$$+ \lambda_9 \Phi_2^\dagger \Delta \Delta^\dagger \Phi_2 + m_\Delta^2 Tr(\Delta^\dagger \Delta) + \bar{\lambda}_8 (Tr \Delta^\dagger \Delta)^2 + \bar{\lambda}_9 Tr(\Delta^\dagger \Delta)^2 \tag{1}$$

We assume that m_{11}^2, m_{22}^2, m_Δ^2, m_{12}^2, $\lambda_{1,2,3,4,5,6,7,8,9}$, $\bar{\lambda}_{8,9}$, $\mu_{1,2,3}$, $\beta_{1,2}$ are real parameters. To avoid tree-level Higgs mediated $FCNC_s$ at tree level, we consider Z_2 symmetry where $\beta_1 = \beta_2 = 0$. Also the Z_2 symmetry is softly broken by the bi-linear terms proportional to m_{12}^2, μ_1, μ_2 and μ_3 parameters. Eleven physical Higgs states occur in the model spectrum: three CP-even neutral Higgs bosons (h_1, h_2, h_3), four simply charged Higgs bosons ($H_1^\pm$, $H_2^\pm$), two CP odd Higgs (A_1, A_2), and finally two doubly charged Higgs bosons $H^{\pm\pm}$. For details see Ref. [35]. The neutral scalar mass matrix reads :

$$\mathcal{M}^2_{\mathcal{CP}_{even}} = \begin{pmatrix} m^2_{\rho_1\rho_1} & m^2_{\rho_2\rho_1} & m^2_{\rho_0\rho_1} \\ m^2_{\rho_1\rho_2} & m^2_{\rho_2\rho_2} & m^2_{\rho_0\rho_2} \\ m^2_{\rho_1\rho_0} & m^2_{\rho_2\rho_0} & m^2_{\rho_0\rho_0} \end{pmatrix} \qquad (2)$$

Its diagonal terms are,

$$m^2_{\rho_1\rho_1} = \lambda_1 v_1^2 + \frac{v_2\left(\sqrt{2}m_3^2 + \mu_3 v_t\right)}{\sqrt{2}v_1}$$

$$m^2_{\rho_2\rho_2} = \lambda_2 v_2^2 + \frac{v_1\left(\sqrt{2}m_3^2 + \mu_3 v_t\right)}{\sqrt{2}v_2}$$

$$m^2_{\rho_0\rho_0} = \frac{4\left(\bar{\lambda}_8 + \bar{\lambda}_9\right)v_t^3 + \sqrt{2}\left(\mu_1 v_1^2 + \mu_3 v_2 v_1 + \mu_2 v_2^2\right)}{2v_t} \qquad (3)$$

while the off-diagonal terms are given by,

$$m^2_{\rho_2\rho_1} = m^2_{\rho_1\rho_2} = \frac{1}{\sqrt{2}}\left(\sqrt{2}v_1 v_2 \lambda_{345} - \sqrt{2}m_3^2 - \mu_3 v_t\right)$$

$$m^2_{\rho_0\rho_1} = m^2_{\rho_1\rho_0} = \frac{1}{\sqrt{2}}\left(\sqrt{2}v_1 v_t (\lambda_6 + \lambda_8) - (2\mu_1 v_1 + \mu_3 v_2)\right)$$

$$m^2_{\rho_0\rho_2} = m^2_{\rho_2\rho_0} = \frac{1}{\sqrt{2}}\left(\sqrt{2}v_2 v_t (\lambda_7 + \lambda_9) - (2\mu_2 v_2 + \mu_3 v_1)\right) \qquad (4)$$

The mass matrix can be diagonalized by an orthogonal matrix $\mathcal{E}$ which can be parametrize as

$$\mathcal{E} = \begin{pmatrix} c_{\alpha_1} c_{\alpha_2} & s_{\alpha_1} c_{\alpha_2} & s_{\alpha_2} \\ -(c_{\alpha_1} s_{\alpha_2} s_{\alpha_3} + s_{\alpha_1} c_{\alpha_3}) & c_{\alpha_1} c_{\alpha_3} - s_{\alpha_1} s_{\alpha_2} s_{\alpha_3} & c_{\alpha_2} s_{\alpha_3} \\ -c_{\alpha_1} s_{\alpha_2} c_{\alpha_3} + s_{\alpha_1} s_{\alpha_3} & -(c_{\alpha_1} s_{\alpha_3} + s_{\alpha_1} s_{\alpha_2} c_{\alpha_3}) & c_{\alpha_2} c_{\alpha_3} \end{pmatrix} \qquad (5)$$

where the mixing angles α_1, α_2 and α_3 can be chosen in the range

$$-\frac{\pi}{2} \leq \alpha_{1,2,3} \leq \frac{\pi}{2} . \qquad (6)$$

the rotation between the two basis (ρ_1, ρ_2, ρ_0) and (h_1, h_2, h_3) diagonalizes the mass matrix $\mathcal{M}^2_{\mathcal{CP}_{even}}$ as,

Table 1 Yukawa couplings of the h_1 boson to the quarks and leptons in 2HDMcT-II

ϕ	ξ_ϕ^u	ξ_ϕ^d	ξ_ϕ^ℓ
h_1	$\frac{s_{\alpha_1} c_{\alpha_2}}{s_\beta}$	$-\frac{c_{\alpha_1} c_{\alpha_2}}{c_\beta}$	$-\frac{c_{\alpha_1} c_{\alpha_2}}{c_\beta}$

$$\mathcal{E} \mathcal{M}^2_{\mathcal{CP}_{even}} \mathcal{E}^T = diag(m_{h_1}^2, m_{h_2}^2, m_{h_3}^2) \tag{7}$$

and leads to three mass eigenstates, ordered by ascending mass as:

$$m_{h_1}^2 < m_{h_2}^2 < m_{h_3}^2 . \tag{8}$$

$\mathcal{L}_{\text{Yukawa}}$ encompasses the entire Yukawa sector of the Two-Higgs Doublet Model (2HDM), along with an additional Yukawa term arising from the triplet field. After spontaneous symmetry breaking, this additional term generates small Majorana mass terms for the neutrinos.

$$- \mathcal{L}_{\text{Yukawa}} \supset -Y_\nu L^T C \otimes i\sigma^2 \Delta L + \text{h.c.} \tag{9}$$

For the type II (2HDMcT-II), up quarks interact with Φ_2, while leptons and down quarks with Φ_1 as :

$$- \mathcal{L}^{II}_{Yukawa} = -y_u \bar{Q}_L \tilde{\Phi}_2 u_R - y_d \bar{Q}_L \Phi_1 d_R - y_\ell \bar{L}_L \Phi_1 \ell_R + h.c , \tag{10}$$

Here Q_L and L_L represent the left-handed quark and lepton doublets, u_R, d_R, and ℓ_R are the right-handed up-type quark, down-type quark, and lepton singlets. Φ_1 and Φ_2 are the two Higgs doublets, with $\tilde{\Phi}_2 = i\sigma_2 \Phi_2^*$. y_u, y_d, and y_ℓ are the Yukawa couplings for the up, down-type quarks and leptons, respectively. The Yukawa couplings of the CP-even Higgs bosons h_1 to the different fermions are presented in Table 1.

3 Results

We first state briefly the set of theoretical and experimental constraints [35, 36] used to perform the phenomenological analysis in 2HDMcT :

- **Unitarity**: The scattering processes must obey perturbative unitarity.
- **Perturbativity**: The quartic couplings of the scalar potential are constrained by the following conditions: $|\lambda_i| < 8\pi$.
- **Vacuum stability**: Boundedness from below (BFB) arising from the positivity in any direction of the fields Φ_i, Δ.
- **Electroweak precision observables**: The oblique parameters S, T and U [37, 38] have been calculated in $2HDMcT$ [36]. The analysis of the precision electroweak data in light of the new PDG mass of the W boson yields [39]:

Table 2 Experimental result of flavor observable: $\bar{B} \to X_s \gamma$ at 95% C.L.

Observable	Experimental result	95% C.L.
BR($\bar{B} \to X_s \gamma$) [36]	$(3.49 \pm 0.19) \times 10^{-4}$ [40]	$[3.11 \times 10^{-4}, 3.87 \times 10^{-4}]$

$$\widehat{S}_0 = -0.01 \pm 0.07, \quad \widehat{T}_0 = 0.04 \pm 0.06, \quad \rho_{ST} = 0.92,$$

We use the following χ^2_{ST} test :

$$\frac{(S - \widehat{S}_0)^2}{\sigma_S^2} + \frac{(T - \widehat{T}_0)^2}{\sigma_T^2} - 2\rho_{ST} \frac{(S - \widehat{S}_0)(T - \widehat{T}_0)}{\sigma_S \sigma_T} \leq R^2 (1 - \rho_{ST}^2), \quad (11)$$

with $R^2 = 2.3$ and 5.99 corresponding to 68.3% and 95% (C.L.s respectively. Our numerical analysis is performed with χ^2_{ST} at 95% C.L.

- The HiggsTools package [27] is employed. This guarantees that the permitted parameter regions are consistent with the observed properties of the 125 GeV Higgs boson (HiggsSignals [27–30]) and with the limits from searches for additional Higgs bosons at the LHC and at LEP (HiggsBounds [27, 31–34]).
- **Flavour constraints**: Flavour constraints are also implemented in our analysis. We used B-physics derived results derived in [36] as well as the experimental data at 2σ [40] displayed in Table 2.

We define the diphoton signal strength $R_{\gamma\gamma}$ as,

$$R_{\gamma\gamma} \equiv \frac{\sigma(gg \to h_1) \times Br(h_1 \to \gamma\gamma)}{\sigma(gg \to H^{SM}) \times Br(H^{SM} \to \gamma\gamma)} \quad (12)$$

Where $\sigma(gg \to h_1)$, $Br(h_1 \to \gamma\gamma)$ and $\sigma(gg \to H^{SM})$, $Br(H^{SM} \to \gamma\gamma)$ are the production cross sections and the branching ratios in 2HDMcT and SM respectively. H^{SM} refer to the observed SM Higgs boson. The subsequent numerical analysis utilizes the following set of input parameters:

$$\mathcal{P}_I = \{\alpha_1, \alpha_2, \alpha_3, m_{h_1}, m_{h_2}, \lambda_1, \lambda_3, \lambda_4, \lambda_6, \lambda_7, \lambda_8, \lambda_9, \bar{\lambda}_8, \bar{\lambda}_9, \mu_1, v_t, \tan\beta\}, \quad (13)$$

with

$$m_{h_1} = 125 \text{ GeV}, \quad m_{h_1} \leq m_{h_2} \leq m_{h_3} \leq 1 \text{ TeV},$$
$$80 \text{ GeV} \leq m_{H_1^\pm}, m_{H_2^\pm}, m_{H^{\pm\pm}} \leq 1 \text{ TeV},$$
$$-2\pi \leq \lambda_1 \leq 2\pi, \quad -2\pi \leq \lambda_{3,4} \leq 2\pi,$$
$$-2\pi \leq \lambda_{6,7} \leq 2\pi, \quad -2\pi \leq \lambda_{8,9} \leq 2\pi, \quad -2\pi \leq \bar{\lambda}_{8,9} \leq 2\pi$$
$$-\pi/2 \leq \alpha_1 \leq \pi/2, \quad \alpha_2 \approx 0, \quad \alpha_3 \approx 0, \quad \mu_1 = v_t = 1 \text{ GeV}, \quad \tan\beta = 8$$
$$(14)$$

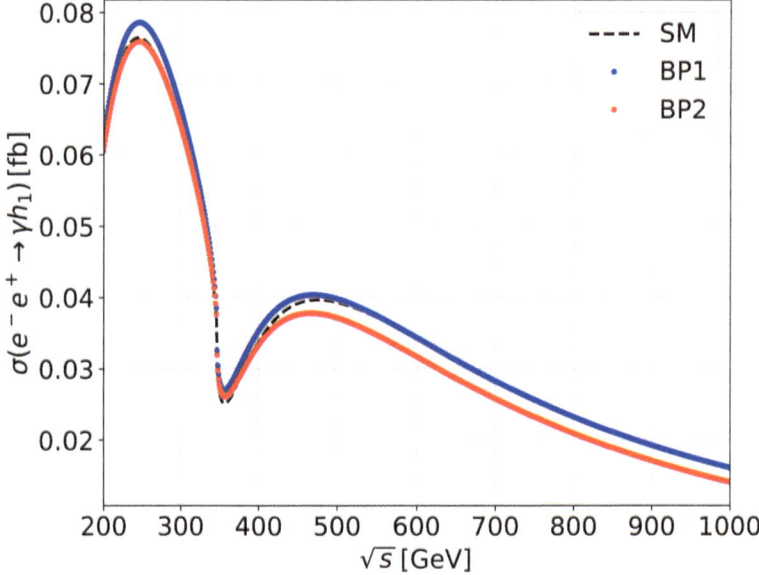

Fig. 1 Total unpolarized cross-section in fb for the process $e^+e^- \to \gamma h_1$ as a function of center-of-mass energy, in two benchmark scenarios. Here we assume that $\alpha_1 = 1.42$ and $m_{h_2} = 673.30$ GeV. The other inputs are given in Table 3

Table 3 Benchmark scenarios used in Fig. 1 where the generated points pass the full set of constraints

Bench.	λ_1	λ_3	λ_4	λ_6	λ_7	λ_8	λ_9	$\bar{\lambda}_8$	$\bar{\lambda}_9$
BP1	1.31	−0.12	1.70	0.64	0.01	0.28	0.38	2.33	0.54
BP2	1.31	0.80	3.06	6.04	2.15	−2.72	0.47	−1.10	3.13

In Fig. 1, we display the unpolarized cross-section of the process $e^-e^+ \to h_1\gamma$ with respect to the SM one as a function of $\sqrt{s}$ in the range 200–1000 GeV. The parameters used in the two Benchmark scenarios BP1 and BP2 are given in Table 3. Our analysis shows that the cross section is very sensitive to the model's parameters and it can exceed its SM value. In Fig. 2 the signal strength denoted as $R_{\gamma\gamma}(h_1)$ and $\sigma(e^-e^+ \to h_1\gamma)$ are depicted as a function of the parameter λ_7 in two distinct scenarios ($\alpha_1 = 1.42, 1.44$). In the upper panels representing the first scenario ($\alpha_1 = 1.42$), the cross section $\sigma(e^-e^+ \to h_1\gamma)$ can reach the SM value, but never quit exceed it because of the BFB conditions. These panels also illustrate that the calculated $R_{\gamma\gamma}$ agrees with the observed signal strength within 1σ C.L, as reported in reference [41] and it is constrained within the range $1.04 \leq R_{\gamma\gamma}(h_1) \leq 1.17$. However, In the lower panels of Fig. 2, representing the second scenario ($\alpha_1 = 1.44$), it is clear that in this case, both $R_{\gamma\gamma}$ and $\sigma(e^-e^+ \to h_1\gamma)$

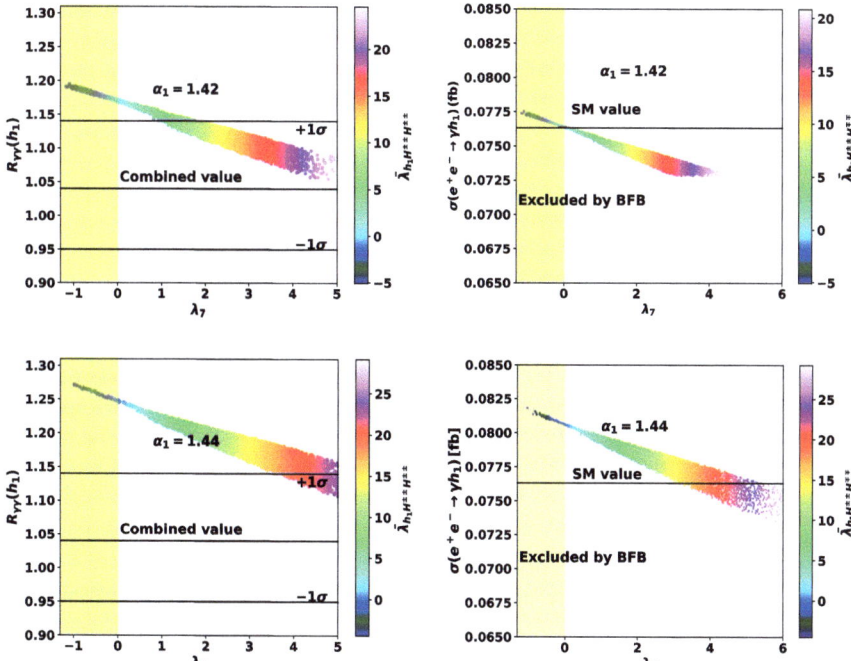

Fig. 2 The signal strength, denoted as $R_{\gamma\gamma}(h_1)$ (left panels) and the cross section for $\sqrt{s} = 250$ GeV (right panels), are depicted as a function of the parameter λ_7 for two values of the parameter α_1, $\alpha_1 = 1.42$ (upper panels), and $\alpha_1 = 1.44$ (lower panels). The plotted points validate all constraints. The color coding shows the variation of the trilinear coupling $\tilde{\lambda}_{h_1 H^{\pm\pm} H^{\mp\mp}}$. The displayed horizontal lines on the left panels denote the central and $\pm 1\sigma$ combined diphoton signal strength reported by ATLAS [41] at 13 TeV. The error for χ^2_{ST} fit is 95.5% C.L. with $\lambda_1 = 1.31$, $\lambda_3 = 6.15$, $\lambda_4 = -3.85$ and $m_{h_2} = 673.30$ GeV, the other inputs are the same as in (14)

undergo a noticeable enhancement. Particularly, $\sigma(e^-e^+ \to h_1\gamma)$ can achieve, and even surpass its SM prediction up to 8.06×10^{-2} fb, while still satisfying BFB conditions. This highlights the remarkable sensitivity of these observables to the model's parameters, especially α_1.

4 Conclusion

In this contribution, we have studied the one-loop process $e^+e^- \to \gamma h_1$ in the framework of 2HDMcT at the e^-e^+ colliders, where the h_1 Higgs boson is chosen to replicate the observed Higgs boson at 125 GeV. Within the parameter space of 2HDMcT, we have shown that the charged new particles in 2HDMcT can alter substantially the cross section as well as the signal strength $R_{\gamma\gamma}$. Our analysis also observed that these observables depend on the model parameters, with a high sensitivity to α_1 in particular. Besides, our results show that the new charged scalars $H_1^\pm$, $H_2^\pm$ and $H^{\pm\pm}$ produce a substantial contribution to both $R_{\gamma\gamma}$ and $\sigma(e^+e^- \to h_1\gamma)$. These contributions leads to $R_{\gamma\gamma}$ that agrees with the observed signal strength within 1σ C.L and allow $\sigma(e^+e^- \to h_1\gamma)$ to reach and even exceed the SM prediction up to 8.06×10^{-2} fb.

References

1. CMS Collaboration, A. Tumasyan et al., A portrait of the Higgs boson by the CMS experiment ten years after the discovery. Nature **607**(7917), 60–68 (2022). http://arxiv.org/abs/2207.00043
2. ATLAS Collaboration, G. Aad et al., A detailed map of Higgs boson interactions by the ATLAS experiment ten years after the discovery. Nature **607**(7917), 52–59 (2022). http://arxiv.org/abs/2207.00092 [Erratum: Nature **612**, E24 (2022)]
3. W. Elmetenawee, Summary of CMS Higgs Physics. http://arxiv.org/abs/2401.07650
4. A. Arbey et al., Physics at the e+ e-linear collider. Eur. Phys. J. C **75**(8), 371 (2015). http://arxiv.org/abs/1504.01726
5. ILC International Development Team Collaboration, A. Aryshev et al., The International Linear Collider: Report to Snowmass 2021. http://arxiv.org/abs/2203.07622
6. P. Bambade et al., The International Linear Collider: A Global Project. http://arxiv.org/abs/1903.01629
7. CLIC Physics Working Group Collaboration, E. Accomando et al., Physics at the CLIC multi-TeV linear collider, in *11th International Conference on Hadron Spectroscopy*, CERN Yellow Reports: Monographs, 6 (2004). http://arxiv.org/abs/hep-ph/0412251
8. CEPC Study Group Collaboration, CEPC Conceptual Design Report: Volume 1 - Accelerator. http://arxiv.org/abs/1809.00285
9. CEPC Study Group Collaboration, M. Dong et al., CEPC Conceptual Design Report: Volume 2 - Physics & Detector. http://arxiv.org/abs/1811.10545
10. CEPC Study Group Collaboration, W. Abdallah et al., CEPC Technical Design Report – Accelerator. http://arxiv.org/abs/2312.14363
11. TLEP Design Study Working Group Collaboration, M. Bicer et al., First look at the physics case of TLEP. J. High Energy Phys. **01**, 164 (2014). http://arxiv.org/abs/1308.6176
12. FCC Collaboration, A. Abada et al., FCC physics opportunities: future circular collider conceptual design report volume 1. Eur. Phys. J. C **79**(6), 474 (2019)
13. R.S. Gupta, H. Rzehak, J.D. Wells, How well do we need to measure the Higgs boson mass and self-coupling? Phys. Rev. D **88**, 055024 (2013) [http://arxiv.org/abs/1305.6397]
14. J. Baglio, C. Weiland, The triple Higgs coupling: a new probe of low-scale seesaw models. J. High Energy Phys. **04**, 038 (2017) [http://arxiv.org/abs/1612.06403]
15. ILC Physics, Detector Study Collaboration, J. Tian, K. Fujii, Measurement of Higgs boson couplings at the International Linear Collider. Nucl. Part. Phys. Proc. **273–275**, 826–833 (2016)
16. C.F. Dürig, Measuring the Higgs Self-coupling at the International Linear Collider. PhD thesis, Hamburg U., Hamburg (2016)
17. T. Liu, K.-F. Lyu, J. Ren, H.X. Zhu, Probing the quartic Higgs boson self-interaction. Phys. Rev. D **98**(9), 093004 (2018) [http://arxiv.org/abs/1803.04359]
18. A. Abbasabadi, D. Bowser-Chao, D.A. Dicus, W.W. Repko, Higgs-boson–photon associated production at eē colliders. Phys. Rev. D **52**, 3919–3928 (1995)
19. A. Djouadi, V. Driesen, W. Hollik, J. Rosiek, Associated production of Higgs bosons and a photon in high-energy e^+e^- collisions. Nucl. Phys. B **491**, 68–102 (1997) [http://arxiv.org/abs/hep-ph/9609420]
20. A. Barroso, J. Pulido, J.C. Romao, Higgs production at e+ e- colliders. Nucl. Phys. B **267**, 509–530 (1986)

21. A. Arhrib, R. Benbrik, T.-C. Yuan, Associated production of Higgs at linear collider in the inert Higgs doublet model. Eur. Phys. J. C **74**, 2892 (2014) [http://arxiv.org/abs/1401.6698]
22. L. Rahili, A. Arhrib, R. Benbrik, Associated production of SM Higgs with a photon in type-II seesaw models at the ILC. Eur. Phys. J. C **79**(11), 940 (2019) [http://arxiv.org/abs/1909.07793]
23. S. Heinemeyer, C. Schappacher, Neutral Higgs boson production at e^+e^- colliders in the complex MSSM: a full one-loop analysis. Eur. Phys. J. C **76**, 220 (2016) [http://arxiv.org/abs/1511.06002]
24. M. Demirci, Associated production of Higgs boson with a photon at electron-positron colliders. Phys. Rev. D **100**(7), 075006 (2019) [http://arxiv.org/abs/1905.09363]
25. ILD Concept Group Collaboration, Y. Aoki, K. Fujii, J. Tian, Study of $e^+e^- \to \gamma h$ at the ILC. http://arxiv.org/abs/2203.07202
26. B.A. Ouazghour, M. Chabab, K. Goure, Higgs photon associated production in a two Higgs doublet type-II seesaw model at future electron-positron colliders. Eur. Phys. J. C **84**(9), 879 (2024) [http://arxiv.org/abs/2403.07722]
27. H. Bahl, T. Biekötter, S. Heinemeyer, C. Li, S. Paasch, G. Weiglein, J. Wittbrodt, HiggsTools: BSM scalar phenomenology with new versions of HiggsBounds and HiggsSignals. Comput. Phys. Commun. **291**, 108803 (2023) [http://arxiv.org/abs/2210.09332]
28. P. Bechtle, S. Heinemeyer, O. Stål, T. Stefaniak, G. Weiglein, *HiggsSignals*: confronting arbitrary Higgs sectors with measurements at the Tevatron and the LHC. Eur. Phys. J. C **74**(2), 2711 (2014) [http://arxiv.org/abs/1305.1933]
29. P. Bechtle, S. Heinemeyer, O. Stål, T. Stefaniak, G. Weiglein, Probing the standard model with Higgs signal rates from the Tevatron, the LHC and a future ILC. J. High Energy Phys. **11**, 039 (2014) [http://arxiv.org/abs/1403.1582]
30. P. Bechtle, S. Heinemeyer, T. Klingl, T. Stefaniak, G. Weiglein, J. Wittbrodt, HiggsSignals-2: probing new physics with precision Higgs measurements in the LHC 13 TeV era. Eur. Phys. J. C **81**(2), 145 (2021) [http://arxiv.org/abs/2012.09197]
31. P. Bechtle, O. Brein, S. Heinemeyer, G. Weiglein, K.E. Williams, HiggsBounds: confronting arbitrary Higgs sectors with exclusion bounds from LEP and the Tevatron. Comput. Phys. Commun. **181**, 138–167 (2010) [http://arxiv.org/abs/0811.4169]
32. P. Bechtle, O. Brein, S. Heinemeyer, G. Weiglein, K.E. Williams, HiggsBounds 2.0.0: confronting neutral and charged Higgs sector predictions with exclusion bounds from LEP and the Tevatron. Comput. Phys. Commun. **182**, 2605–2631 (2011) [http://arxiv.org/abs/1102.1898]
33. P. Bechtle, O. Brein, S. Heinemeyer, O. Stål, T. Stefaniak, G. Weiglein, K.E. Williams, HiggsBounds − 4: improved tests of extended Higgs Sectors against exclusion bounds from LEP, the Tevatron and the LHC. Eur. Phys. J. C **74**(3), 2693 (2014) [http://arxiv.org/abs/1311.0055]
34. P. Bechtle, D. Dercks, S. Heinemeyer, T. Klingl, T. Stefaniak, G. Weiglein, J. Wittbrodt, HiggsBounds-5: testing Higgs sectors in the LHC 13 TeV era. Eur. Phys. J. C **80**(12), 1211 (2020) [http://arxiv.org/abs/2006.06007]
35. B. Ait Ouazghour, A. Arhrib, R. Benbrik, M. Chabab, L. Rahili, Theory and phenomenology of a two-Higgs-doublet type-II seesaw model at the LHC run 2. Phys. Rev. D **100**(3), 035031 (2019) [http://arxiv.org/abs/1812.07719]
36. B. Ait Ouazghour, M. Chabab, The two Higgs doublet type-II seesaw model: naturalness and $B^- \to X_s\gamma$ versus heavy Higgs masses. Phys. Lett. B **846**, 138241 (2023). [http://arxiv.org/abs/2305.08030]
37. M.E. Peskin, T. Takeuchi, Estimation of oblique electroweak corrections. Phys. Rev. **D46**, 381–409 (1992)
38. W. Grimus, L. Lavoura, O.M. Ogreid, P. Osland, The Oblique parameters in multi-Higgs-doublet models. Nucl. Phys. B **801**, 81–96 (2008) [http://arxiv.org/abs/0802.4353]
39. Particle Data Group Collaboration, R.L. Workman et al., Review of particle physics. Prog. Theor. Exp. Phys. **2022**, 083C01 (2022)

40. HFLAV Collaboration, Y.S. Amhis et al., Averages of b-hadron, c-hadron, and τ-lepton properties as of 2021. Phys. Rev. D **107**, 052008 (2023) [http://arxiv.org/abs/2206.07501]
41. ATLAS Collaboration, G. Aad et al., Measurement of the properties of Higgs boson production at $\sqrt{s} = 13$ TeV in the $H \to \gamma\gamma$ channel using 139 fb^{-1} of pp collision data with the ATLAS experiment. J. High Energy Phys. **07**, 088 (2023) [http://arxiv.org/abs/2207.00348]

Open Access This chapter is licensed under the terms of the Creative Commons Attribution 4.0 International License (http://creativecommons.org/licenses/by/4.0/), which permits use, sharing, adaptation, distribution and reproduction in any medium or format, as long as you give appropriate credit to the original author(s) and the source, provide a link to the Creative Commons license and indicate if changes were made.

The images or other third party material in this chapter are included in the chapter's Creative Commons license, unless indicated otherwise in a credit line to the material. If material is not included in the chapter's Creative Commons license and your intended use is not permitted by statutory regulation or exceeds the permitted use, you will need to obtain permission directly from the copyright holder.

Collider Constraints on Massive Gravitons

Malak Ait Tamlihat, David d'Enterria, Laurent Schoeffel, Hua-Sheng Shao, and Yahya Tayalati

1 Introduction

General Relativity (GR) serves as a classical field theory describing gravity through an interacting massless tensor (spin-2) field. This theory, when quantized, introduces gravitons, which are considered massless due to the diffeomorphism invariance of GR. However, gauge invariance doesn't universally guarantee zero masses for gauge states, and quantum effects from other fields can confer mass to gravitons without compromising the fundamental properties of GR [1]. Although the existence of a self-consistent quantum field theory of GR applicable across all energy scales remains an open question, employing an Effective Field Theory (EFT) approach enables the exploration of practical consequences. This approach has revealed the emergence of massive spin-2 degrees of freedom in various gravity modifications.

M. Ait Tamlihat (✉)
Faculty of Sciences, Mohammed V University in Rabat, Rabat, Morocco
e-mail: malak.ait.tamlihat@cern.ch

D. d'Enterria
CERN, EP Department, Geneva, Switzerland

L. Schoeffel
Irfu, CEA, Université Paris-Saclay, Gif-sur-Yvette, France

H.-S. Shao
Laboratoire de Physique Théorique et Hautes Energies (LPTHE), UMR 7589, Sorbonne Université et CNRS, Paris Cedex 05, France

Y. Tayalati
Faculty of Sciences, Mohammed V University in Rabat, Rabat, Morocco

Institute of Applied Physics, Mohammed VI Polytechnic University, Hay Moulay Rachid, Ben Guerir, Morocco

© The Author(s) 2026
Y. Tayalati, M. Gouighri (eds.), *The First African Conference on High Energy Physics*, Springer Proceedings in Physics 425,
https://doi.org/10.1007/978-3-031-88933-2_24

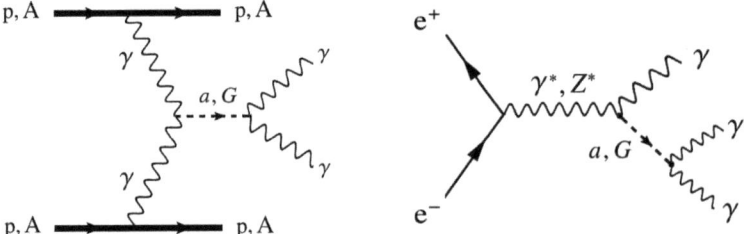

Fig. 1 Diagrams showing the decay of an ALP or graviton into two photons (left), and the production of an ALP or graviton in e^+e^- collisions, leading to a triphoton final state (right)

The interaction between the graviton and photons [2] is described within this context, by the following formula:

$$L_\gamma^G = g_{G\gamma}\left(-F_{\mu\rho}F_\nu^\rho + \frac{1}{4}\eta_{\mu\nu}(F_{\rho\sigma})^2\right)G^{\mu\nu}, \qquad (1)$$

where $F_{\mu\rho}$ is the electromagnetic field, $\eta_{\mu\nu}$ the flat spacetime metric, and $g_{G\gamma}$ is the G-γ coupling. Photon-photon collisions offer exceptional environment for probing particles beyond the standard model (BSM) that interact with photons, such as spin-0 Axion-Like Particles (ALPs) and spin-2 tensor particles like gravitons. In this study, we establish upper bounds on $g_{G\gamma}$ across a range of graviton masses m_G spanning from 100 MeV to 2 TeV. Through simulating various productions of both ALPs and gravitons, based on the digarms depicted in Fig. 1 and considering the distinctive cross sections, $\gamma\gamma$ partial widths, and decay characteristics between both particles, we reinterpret the existing experimental limits on the ALP-γ coupling to deduce constraints for the G-γ coupling. We derive these constraints in two distinct contexts. Initially, we adopt a simplified approach assuming a 100% decay branching fraction of the graviton into two photons ($\mathcal{B}_{G\to\gamma\gamma} = 1$). This scenario, commonly employed in ALPs searches [3–10], maximizes the sensitivity of the graviton-photon coupling. Subsequently, we examine a more practical situation in which the graviton has universal couplings to all Standard Model particles. In this case, the graviton predominantly decays into diphotons only for lower m_G values. However, as we go beyond a few GeV, the emergence of kinematic phase space for decays to massive Standard Model fermions or bosons becomes significant. Consequently, the branching fraction decreases to $\mathcal{B}_{G\to\gamma\gamma} \approx 0.05$. The universal coupling scenario enables a proper computation of the $e^+e^- \to G\gamma$ cross sections without encountering issues related to the violation of perturbative unitarity.

2 Results

When considering graviton limits from PbPb or pp Ultra-Peripheral Collisions (UPCs), it is possible to maintain the simplifying assumption of unity diphoton-decay branching fractions ($\mathcal{B}_{G\to\gamma\gamma} = 1$). The corresponding upper exclusion limits

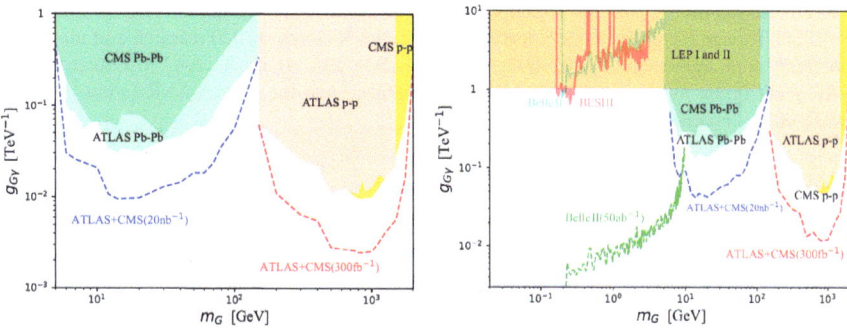

Fig. 2 Exclusion Limits at 95% CL on graviton-photon coupling, with Br(G $\to \gamma\gamma$) = 1 (Left) and an universal graviton coupling to SM particles (Right). Projected limits (indicated by dashed lines) are also showcased for the ultimate integrated luminosities anticipated at LHC and Belle II

at a 95% confidence level (CL) for the graviton-photon coupling ($g_{G\gamma}$) are illustrated in the left of Fig. 2. Additionally, we present extrapolated limits (depicted by dashed curves) by projecting the current findings to the future integrated luminosities in future PbPb and pp collisions at the HL-LHC. Specifically, we consider $\mathcal{L} = 20$ nb^{-1} for PbPb and a conservative estimate of $\mathcal{L} = 300$ fb^{-1} for the pp scenario. To derive $\mathcal{B}_{G\to\gamma\gamma}$ limits at e^+e^- colliders, we take into account the universal coupling scenario. Within this framework, the diphoton branching ratio of the graviton is set for each m_G. The results are depicted in Fig. 2 (Right), showcasing upper limits on the graviton-photon coupling across the range of $g_{G\gamma} \approx 1$–0.05 TeV^{-1} for masses $m_G \approx 100$ MeV–2 TeV. Additionally, the figure illustrates extrapolated limits (dashed curves) corresponding to the anticipated total integrated luminosities throughout the entire operational lifetime of the HL-LHC and Belle II experiments. These extrapolations suggest the potential for improving current constraints by approximately 100-fold in the low-mass region and by a factor of 4 at higher masses.

The findings showcased in this study rival existing inclusive graviton searches. The exclusive production and decay of gravitons into photon pairs ($\gamma\gamma$) are subject to negligible standard model backgrounds. Through meticulous control of instrumental effects and consistent integrated luminosities, the coupling limits for G-γ derived from exclusive analyses have the potential to surpass those obtained from standard inclusive graviton searches at the LHC.

3 Conclusion

In this work, we explored the prospect of detecting massive spin-2 gravitons coupling to photons. We established 95% upper limits for $g_{G\gamma}$ in the range of approximately 1–0.1 TeV^{-1} for graviton masses $m_G = 5$–100 GeV in PbPb and pp collisions, and over $g_{G\gamma} \approx 0.5$–0.05 TeV^{-1} for $m_G = 150$ GeV–2 TeV.

Acknowledgments This work is supported by the Moroccan Ministry of Higher Education, Scientific Research and Innovation, the European Union's Horizon 2020 research and innovation program (grant agreement No. 824093, STRONG-2020, EU Virtual Access "NLOAccess"), the ERC grant (grant agreement ID 101041109, "BOSON") and the French ANR (grant ANR-20-CE31-0015, "PrecisOnium").

References

1. K. Hinterbichler, Theoretical aspects of massive gravity. Rev. Mod. Phys. **84**(3), 671–710 (2012). https://doi.org/10.1103/RevModPhys.84.671
2. M. Fierz, W. Pauli, On relativistic wave equations for particles of arbitrary spin in an electromagnetic field. Proc. R. Soc. Lond. A **173**, 211–232 (1939). https://doi.org/10.1098/rspa.1939.0140
3. ATLAS Collaboration, G. Aad et al., Measurement of light-by-light scattering and search for axion-like particles with 2.2 nb^{-1} of Pb+Pb data with the ATLAS detector. J. High Energy Phys. **2021**(03), 243 (2021). https://doi.org/10.1007/JHEP11(2021)050
4. CMS Collaboration, A.M. Sirunyan et al., Evidence for light-by-light scattering and searches for axion-like particles in ultraperipheral PbPb collisions at $\sqrt{s_{NN}} = 5.02$ TeV. Phys. Lett. B **797**, 134826 (2019). https://doi.org/10.1016/j.physletb.2019.134826
5. ATLAS Collaboration, G. Aad et al., Search for an axion-like particle with forward proton scattering in association with photon pairs at ATLAS. J. High Energy Phys. **2023**(07), 234 (2023). https://doi.org/10.1007/JHEP07(2023)234
6. TOTEM and CMS Collaboration, A. Tumasyan et al., First search for exclusive diphoton production at high mass with tagged protons in proton-proton collisions at $\sqrt{s} = 13$ TeV. Phys. Rev. Lett. **129**, 011801 (2022). https://doi.org/10.1103/PhysRevLett.129.011801
7. Belle-II Collaboration, F. Abudinén et al., Search for axion-like particles produced in e^+e^- collisions at Belle II. Phys. Rev. Lett. **125**, 161806 (2020). https://doi.org/10.1103/PhysRevLett.125.161806
8. BESIII Collaboration, M. Ablikim et al., Search for an axion-like particle in radiative J/ψ decays. Phys. Lett. B **838**, 137698 (2023). https://doi.org/10.1016/j.physletb.2023.137698
9. M. Baillargeon, F. Boudjema, E. Chopin, V. Lafage, New physics with three photon events at LEP. Z. Phys. C **71**, 431–442 (1996). https://doi.org/10.1007/s002880050188
10. D. d'Enterria, M. Ait Tamlihat, L. Schoeffel, H. Shao, Y. Tayalati, Collider constraints on massive gravitons coupling to photons. Phys. Lett. B **846**, 138237 (2023). https://doi.org/10.1016/j.physletb.2023.138237

Open Access This chapter is licensed under the terms of the Creative Commons Attribution 4.0 International License (http://creativecommons.org/licenses/by/4.0/), which permits use, sharing, adaptation, distribution and reproduction in any medium or format, as long as you give appropriate credit to the original author(s) and the source, provide a link to the Creative Commons license and indicate if changes were made.

The images or other third party material in this chapter are included in the chapter's Creative Commons license, unless indicated otherwise in a credit line to the material. If material is not included in the chapter's Creative Commons license and your intended use is not permitted by statutory regulation or exceeds the permitted use, you will need to obtain permission directly from the copyright holder.

Results From the NA62 Experiment at CERN

Mattia Soldani

1 Introduction

Over the last seven decades, measurements of kaon decays have played a central role in both testing the Standard Model (SM) and searching for New Physics (NP). Indeed, the kaon sector proves a powerful probe at the intensity frontier, owing to a small number of decay modes, to rather simple final states and to the accessibility of intense kaon beams [7].

The NA62 experiment is currently carrying on the successful series of kaon decay experiments at the CERN North Area [4]. It makes use of a 400 GeV/c primary proton beam extracted from the CERN Super Proton Synchrotron with $(1.9–2.2) \times 10^{12}$ protons per (~4.8 s long) pulse and a relative duty cycle of

---

Mattia Soldani, on behalf of the NA62 Collaboration: A. Akmete, R. Aliberti, F. Ambrosino, R. Ammendola, B. Angelucci, A. Antonelli, G. Anzivino, R. Arcidiacono, T. Bache, A. Baeva, D. Baigarashev, L. Bandiera, M. Barbanera, J. Bernhard, A. Biagioni, L. Bician, C. Biino, A. Bizzeti, T. Blazek, B. Bloch-Devaux, P. Boboc, V. Bonaiuto, M. Boretto, M. Bragadireanu, A. Briano Olvera, D. Britton, F. Brizioli, M.B. Brunetti, D. Bryman, F. Bucci, T. Capussela, J. Carmignani, A. Ceccucci, P. Cenci, V. Cerny, C. Cerri, B. Checcucci, A. Conovaloff, P. Cooper, E. Cortina Gil, M. Corvino, F. Costantini, A. Cotta Ramusino, D. Coward, P. Cretaro, G. D'Agostini, J. Dainton, P. Dalpiaz, H. Danielsson, M. D'Errico, N. De Simone, D. Di Filippo, L. Di Lella, N. Doble, B. Dobrich, F. Duval, V. Duk, D. Emelyanov, J. Engelfried, T. Enik, N. Estrada-Tristan, V. Falaleev, R. Fantechi, V. Fascianelli, L. Federici, S. Fedotov, A. Filippi, R. Fiorenza, M. Fiorini, O. Frezza, J. Fry, J. Fu, A. Fucci, L. Fulton, E. Gamberini, L. Gatignon, G. Georgiev, S. Ghinescu, A. Gianoli, M. Giorgi, S. Giudici, F. Gonnella, K. Gorshanov, E. Goudzovski, C. Graham, R. Guida, E. Gushchin, F. Hahn, H. Heath, J. Henshaw, Z. Hives, E.B. Holzer, T. Husek, O. Hutanu, D. Hutchcroft, L. Iacobuzio, E. Iacopini, E. Imbergamo, B. Jenninger, J. Jerhot, R.W. Jones, K.

M. Soldani (✉)
INFN Laboratori Nazionali di Frascati, Frascati, Italy

∼0.3 [7]. The primary beam impinges on a beryllium target, thus producing a secondary, unseparated positive hadron beam, with a total rate of 750 MHz consisting of approximately 70% charged pions, 23% protons and 6% charged kaons. The 75 GeV ± 1% momentum component of the secondary beam is used by the experiment [7].

The NA62 apparatus essentially consists of an instrumented 117 m long vacuum tank, with variable diameter between 1.92 m (upstream) and 2.8 m (downstream). It was designed with a strong focus on excellent time resolution for matching the incident kaon and the decay output tracks with a precision of $\mathcal{O}(100$ ps); a powerful and redundant charged-particle identification (PID) system; a highly hermetic coverage of the decay volume, to guarantee that the full reconstruction of both the signal kinematics and the background veto (especially the channels with final-state photons from π^0 decays) is performed with high efficiency [7].

Firstly, the kaon component of the secondary beam is identified with high efficiency ($\gtrsim$99%) by the KTAG (Kaon TAGger) [22], a CEDAR (ChErenkov Differential counter with Achromatic Ring focus) derived from the CERN CEDAR-W design with custom photodetection and readout systems. The track and momentum of the beam particle is also measured with a beam spectrometer, the Giga TracKer (GTK) [27], which consists of an array of silicon pixel modules installed around 4 dipole magnets arranged as a magnetic achromat. Each module is 510 μm thick, with a 300 × 300 μm^2 pixel pitch. Overall, an angular resolution of 16 μrad and a relative momentum resolution of 0.2% are attained.

The fiducial volume (FV), i.e., the region in which the kaon primary decay vertices are sought, occupies the first 60 m of the vacuum tank, downstream of the GTK. Only about 4.5 MHz of charged kaons, i.e., 10% of those entering the experiment, decay in the FV. Several subsystems are devoted to the measurement of the decay products. A brief summary of the main elements of the apparatus is provided in the following. Further details can be found in Ref. [7].

Kampf, V. Kekelidze, D. Kereibay, S. Kholodenko, G. Khoriauli, A. Khotyantsev, A. Kleimenova, A. Korotkova, M. Koval, V. Kozhuharov, Z. Kucerova, Y. Kudenko, J. Kunze, V. Kurochka, V. Kurshetsov, G. Lanfranchi, G. Lamanna, E. Lari, G. Latino, P. Laycock, C. Lazzeroni, M. Lenti, G. Lehmann Miotto, E. Leonardi, P. Lichard, L. Litov, P. Lo Chiatto, R. Lollini, D. Lomidze, A. Lonardo, P. Lubrano, M. Lupi, N. Lurkin, D. Madigozhin, I. Mannelli, A. Mapelli, F. Marchetto, R. Marchevski, S. Martellotti, P. Massarotti, K. Massri, E. Maurice, A. Mazzolari, M. Medvedeva, A. Mefodev, E. Menichetti, E. Migliore, E. Minucci, M. Mirra, M. Misheva, N. Molokanova, M. Moulson, S. Movchan, M. Napolitano, I. Neri, F. Newson, A. Norton, M. Noy, T. Numao, V. Obraztsov, A. Okhotnikov, A. Ostankov, S. Padolski, R. Page, V. Palladino, I. Panichi, A. Parenti, C. Parkinson, E. Pedreschi, M. Pepe, M. Perrin-Terrin, L. Peruzzo, D. Petrov, Y. Petrov, F. Petrucci, R. Piandani, M. Piccini, J. Pinzino, I. Polenkevich, L. Pontisso, Yu. Potrebenikov, D. Protopopescu, M. Raggi, M. Reyes Santos, M. Romagnoni, A. Romano, P. Rubin, G. Ruggiero, V. Ryjov, A. Sadovsky, A. Salamon, C. Santoni, G. Saracino, F. Sargeni, S. Schuchmann, V. Semenov, A. Sergi, A. Shaikhiev, S. Shkarovskiy, M. Soldani, D. Soldi, M. Sozzi, T. Spadaro, F. Spinella, A. Sturgess, V. Sugonyaev, J. Swallow, A. Sytov, G. Tinti, A. Tomczak, S. Trilov, M. Turisini, P. Valente, B. Velghe, S. Venditti, P. Vicini, R. Volpe.

The charged particle trajectories are reconstructed by a magnetic spectrometer, which includes a bending magnet [21] (with a transverse momentum kick of 270 MeV/c) instrumented by 4 straw-tube modules [7] (STRAW) with a spatial resolution of 130 μm and a momentum resolution of $0.3\% + 0.005\% p$ (p being the particle momentum), together with a Charged-particle HODoscope (CHOD) [25].

A hermetic photon-veto system provides efficient coverage to the detection of photons from several background sources in a very wide angular range. In particular, the Large-Angle Veto (LAV) [1] encompasses 12 modules of ring-shaped arrays of lead glass blocks, deployed around and downstream of the FV, providing an angular coverage up to 50 mrad. On the other hand, angles ≤ 8.5 mrad are covered by electromagnetic calorimeters placed at the downstream end of the apparatus: the NA48 Liquid Krypton (LKr) calorimeter [20], the Intermediate-Ring Calorimeter (IRC) and the Small-Angle Calorimeter (SAC) [7]. Overall, a π^0 suppression at the level of 10^{-8} is achieved.

Hadronic calorimeters [7, 20] and a Ring-Image Cherenkov Detector (RICH) [2, 3] allow for a powerful PID. Overall, a muon suppression at the level of 10^{-7} is achieved. Moreover, the RICH provides a high-time-resolution (<100 ps) reference for the output state [3] and allows for additional muon-pion separation at the level of 10^{-2}.

2 $K^+ \to \pi^+ \nu \bar{\nu}$ and $K^+ \to \pi^+ X$

The $K^+ \to \pi^+ \nu \bar{\nu}$ (PNN) decay (as well as its neutral counterpart, $K_L \to \pi^0 \nu \bar{\nu}$) is a flavor-changing neutral current process that proceeds via Z-penguin and double-W-box diagrams [7]. It is heavily suppressed because of the GIM mechanism and of the CKM hierarchy, with a branching ratio $BR_{PNN}^{SM} = (7.86 \pm 0.61) \times 10^{-11}$ in the SM [18]. See also, e.g., [5, 6]. As the decay dynamics is dominated by short-distance contributions and its hadronic matrix element can be obtained from the experimental measurement of the K_{e3} channel ($K^+ \to \pi^0 e^+ \nu_e$) with sub-% precision, rather clean theoretical predictions are accessible in the SM. The $K^+ \to \pi^+ \nu \bar{\nu}$ decay may prove highly sensitive to flavor NP, including, e.g., non-SM behaviour due to new sources of flavor violation, direct CP violation, lepton-flavor non-universality and leptoquarks [7].

Essentially, the PNN reconstruction proceeds by identifying an input K^+ track that matches an output π^+ track with some missing tranverse momentum (due to the neutrinos) with the decay vertex inside the FV. In the (p_π, m_{miss}^2) plane two boxes are identified as the signal region (Fig. 1 left) and several other regions are defined for validation and evaluation of three of the main background sources, i.e., $K^+ \to \mu^+ \nu_\mu$, $K^+ \to \pi^+ \pi^+ \pi^-$ and $K^+ \to \pi^+ \pi^0$; the latter is also used as a normalization channel. Two other contributions to the overall background are noteworthy: the one from $K^+ \to \pi^+ \pi^- e^+ \nu_e$, which only affects the signal box at higher m_{miss}^2 and was characterized with simulations only, and the so-called upstream background, given by π^+ coming from K^+ interactions or decay in flight

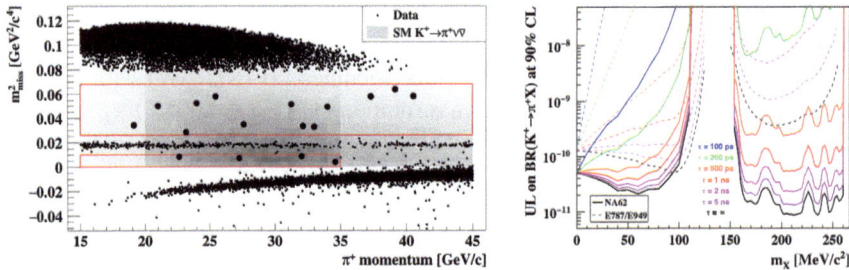

Fig. 1 *Left:* PNN candidate events from the 2018 dataset in the $(p_\pi, m_{\text{miss}}^2)$ plane. The intensity of the gray shaded area reflects the variation of the SM signal acceptance. The signal regions are highlighted in red. *Right:* Model-independent observed BR upper limits for different X mass and lifetime hypotheses. Reprinted under CC-BY-4.0 license from Ref. [10]. © 2021, CERN for the benefit of the NA62 Collaboration

occurring upstream of the FV, which was characterized with a data-driven approach. The analysis procedure is described in extensive detail in Refs. [8–10].

The datasets collected in 2016 [8], 2017 [9] and 2018 [10] (in Fig. 1 left) have been analyzed. Overall, a single-event sensitivity of $(0.839 \pm 0.053_{\text{syst}}) \times 10^{-11}$ has been reached [10]. 20 candidate events have been observed in the signal region, to be compared to the expected sum of $10.01 \pm 0.42_{\text{syst}} \pm 1.19_{\text{ext}}$ PNN events and $7.03^{+1.05}_{-0.82}$ background events [10]. This leads to $\text{BR}_{\text{PNN}}^{\text{meas}} = (10.6^{+4.0}_{-3.4}|_{\text{stat}} \pm 0.9_{\text{syst}}) \times 10^{-11}$ at 68% CL, which is the most precise measurement to date, and corresponds to 3.4σ evidence for the existence of the decay channel [10].

The event sample selected in the PNN analysis has been used to search for evidence of the $K^+ \to \pi^+ X$ decay, where X is a dark scalar or pseudo-scalar. X can be stable, decay to other invisible particles or live long enough to decay outside the NA62 apparatus. This search proceeds under the assumption that all the events observed in the PNN signal regions correspond to the expected background— i.e., including the PNN events, which in this case constitute the main background source. As shown in Fig. 1 right, upper limits are established on $\text{BR}(K^+ \to \pi^+ X)$ for different X mass and lifetime hypotheses. Further details can be found in Ref. [10].

3 $K^+ \to \pi^+ e^+ e^- e^+ e^-$

In the SM, the $K^+ \to \pi^+ e^+ e^- e^+ e^-$ decay proceeds mainly through the $\pi^+ \pi^0$ state followed by the double-Dalitz $\pi^0 \to e^+ e^- e^+ e^-$ decay ("$K_{2\pi\text{DD}}$"), which is resonant and has a branching ratio $\text{BR}_{2\pi\text{DD}}^{\text{meas}} = (6.9 \pm 0.3) \times 10^{-6}$ [23]. This final state also features a non-resonant contribution ("$K_{\pi 4e}$"), given by an intermediate state with a one-/two-photon exchange and with a branching ratio $\text{BR}_{2\pi\text{DD}}^{\text{theo}} = (7.2 \pm 0.7) \times 10^{-11}$ in the SM [17].

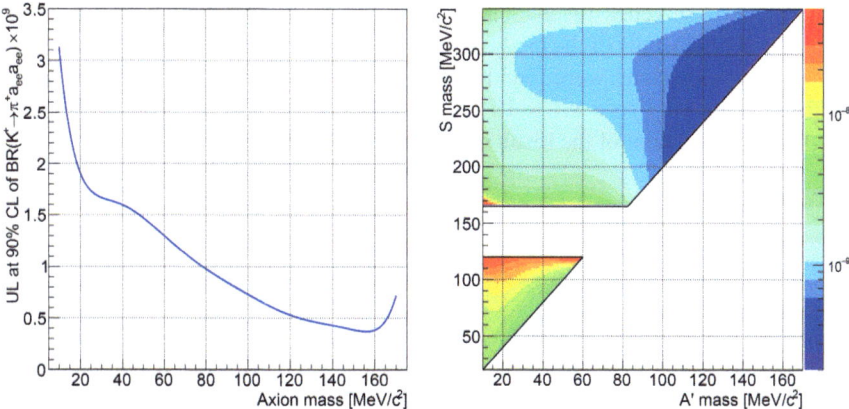

Fig. 2 Upper limits at 90% CL of the decay chains $K^+ \to \pi^+ aa \to \pi^+ e^+ e^+ e^- e^-$ (left) and $K^+ \to \pi^+ S \to \pi^+ A' A' \to \pi^+ e^+ e^+ e^- e^-$ (right) as a function of the masses of the dark mediators involved. Reprinted under CC-BY-4.0 license from [17]. © 2023, The Author(s)

The $K_{\pi 4e}$ channel is also a suitable probe for the dark sector. In particular, the decay might proceed through the $K^+ \to \pi^+ aa$ state, where a is a dark axion decaying into an electron-positron pair. Moreover, a scenario involving an intermediate state with a dark scalar S that decays into two dark photons A', which in turn decay into an electron-positron pair each, is considered.

The rather characteristic kinematics of the final state allows for this analysis to rely exclusively on the STRAW data, which prevents an approximately tenfold loss of signal acceptance. Omitting the information from the other detectors in the apparatus reduces the selectivity, which is feasible because strong kinematic constraints can be used. The $K_{2\pi\mathrm{DD}}$ channel is used for normalization, whereas the signal events are sought in the non-resonant part. Extensive details of the event selection procedure are provided in Ref. [17].

The analysis was performed on the 2017–2018 data. No signal candidates were observed within the signal selection cuts [17]. This corresponds to the upper limit $\mathrm{BR}^{\mathrm{meas}}_{\pi 4e} < 1.4 \times 10^{-8}$ at 90% CL, which is ~200 times larger than the SM expectation [17]. Upper limits were also estimated for decay chains initiating from axions and scalar dark-sector particles, as a function of the masses of the dark mediators involved: they are shown in Fig. 2. In particular, the QCD axion model is excluded as a possible explanation of the $X17$ observations [26].

4 $K^+ \to \mu^- \nu e^+ e^+$

Depending on the flavor of the neutrino, the $K^+ \to \mu^- \nu e^+ e^+$ ("$K_{\mu\nu ee}$") decay violates conservation of either lepton flavour or both lepton flavour and lepton number. NA62 searched for evidence of this decay in the data collected between 2016 and 2018. Details of the analysis can be found in Ref. [14].

No signal candidates were found within the signal selection box, which leads to the upper limit $\text{BR}^{\text{meas}}_{\mu\nu ee} < 8.1 \times 10^{-11}$ at 90% CL [14]. This also corresponds to a 250-fold improvement of the result of previous searches [19].

5 $A' \to \mu^+\mu^-$ in Beam-Dump Mode

A search for evidence of the $A' \to \mu^+\mu^-$ decay, where A' is a dark photon with a mass of $\mathcal{O}(100 \text{ MeV})$, can be made by running the NA62 experiment in dump mode. In this mode, which is described in detail in Ref. [16], the 400 GeV/c protons from the Super Proton Synchrotron directly impinge on the fully-closed hadron beam collimators (TAXes—800 mm of copper and 2400 mm of iron).

About 10 days of dump-mode run were performed in 2021. The resulting data were used to search for the aforementioned decay, as discussed in Ref. [16]. One event passed the signal selection criteria, which would correspond to a 2.4σ global significance [16]. This event could be interpreted as background, as it is located close to the border of the signal selection cuts; indeed, the probability of background observation within the signal region is 1.6% [16]. Overall, no evidence of the decay under study is established [16].

6 Conclusions and Outlook

The NA62 experiment is currently very active in exploring the kaon sector. Firstly, the PNN decay channel, for which an evidence at the level of 3.4σ was reached, is being thoroughly characterized. Moreover, numerous searches for NP through dark mediators, such as in $K^+ \to \pi^+ X$, in $K^+ \to \pi^+ e^+ e^- e^+ e^-$ and in $A' \to \mu^+\mu^-$ (measured in dump mode), and lepton-flavour/number-violating processes like $K^+ \to \mu^- \nu e^+ e^+$ are being performed. Other searches done by the NA62 experiment are discussed in Refs. [12, 13, 15]. No evidence of deviations from the SM have been found so far.

At the same time, the HIKE (High Intensity Kaon Experiments) experiment [11, 24] is under development, with the aim of succeeding NA62 after the end of its operations, expected in late 2025, bringing charged-kaon measurements to an unprecedented sensitivity level and extending the kaon program at CERN to K_L physics.

References

1. A. Antonelli et al., Performance of the NA62 LAV front-end electronics. J. Instrum. **8**(01), C01020 (2013). https://doi.org/10.1088/1748-0221/8/01/C01020
2. G. Anzivino et al., Precise mirror alignment and basic performance of the RICH detector of the NA62 experiment at CERN. J. Instrum. **13**(07), P07012 (2018). https://doi.org/10.1088/1748-0221/13/07/P07012
3. G. Anzivino et al., Light detection system and time resolution of the NA62 RICH. J. Instrum. **15**(10), P10025 (2020). https://doi.org/10.1088/1748-0221/15/10/P10025
4. D. Banerjee et al., The North Experimental Area at the CERN Super Proton Synchrotron. Tech. Rep. CERN-ACC-NOTE-2021-0015, CERN (2021). https://cds.cern.ch/record/2774716
5. J. Brod, M. Gorbahn, E. Stamou, Updated standard model prediction for $K \to \pi \nu \bar{\nu}$ and ϵ_K. Proc. Sci. **BEAUTY2020**, 056 (2021). https://doi.org/10.22323/1.391.0056
6. A.J. Buras, Standard model predictions for rare K and B decays without new physics infection. Eur. Phys. J. C **83**(1), 66 (2023). https://doi.org/10.1140/epjc/s10052-023-11222-6
7. E. Cortina Gil et al., The beam and detector of the NA62 experiment at CERN. J. Instrum. **12**(05), P05025 (2017). https://doi.org/10.1088/1748-0221/12/05/P05025
8. E. Cortina Gil et al., First search for $K^+ \to \pi^+ \nu \bar{\nu}$ using the decay-in-flight technique. Phys. Lett. B **791**, 156–166 (2019). https://doi.org/10.1016/j.physletb.2019.01.067
9. E. Cortina Gil et al., An investigation of the very rare $K^+ \to \pi^+ \nu \bar{\nu}$ decay. J. High Energy Phys. **2011**, 042 (2020). https://doi.org/10.1007/JHEP11(2020)042
10. E. Cortina Gil et al., Measurement of the very rare $K^+ \to \pi^+ \nu \bar{\nu}$ decay. J. High Energy Phys. **06**, 093 (2021). https://doi.org/10.1007/JHEP06(2021)093
11. E. Cortina Gil et al., HIKE, High Intensity Kaon Experiments at the CERN SPS: Letter of Intent. Tech. Rep. CERN-SPSC-2022-031, SPSC-I-257, SPSC-I-257, CERN, Geneva (2022). https://cds.cern.ch/record/2839661
12. E. Cortina Gil et al., Searches for lepton number violating $K^+ \to \pi^-(\pi^0)e^+e^+$ decays. Phys. Lett. B **830**, 137172 (2022). https://doi.org/10.1016/j.physletb.2022.137172
13. E. Cortina Gil et al., A measurement of the $K^+ \to \pi^+ \mu^+ \mu^-$ decay. J. High Energy Phys. **11**, 011 (2022). https://doi.org/10.1007/JHEP06(2023)040
14. E. Cortina Gil et al., A search for the $K^+ \to \mu^- \nu e^+ e^+$ decay. Phys. Lett. B **838**, 137679 (2023). https://doi.org/10.1016/j.physletb.2023.137679
15. E. Cortina Gil et al., A study of the $K^+ \to \pi^0 e^+ \nu \gamma$ decay. J. High Energy Phys. **09**, 040 (2023). https://doi.org/10.1007/JHEP09(2023)040
16. E. Cortina Gil et al., Search for dark photon decays to $\mu^+ \mu^-$ at NA62. J. High Energy Phys. **09**, 035 (2023). https://doi.org/10.1007/JHEP09(2023)035
17. E. Cortina Gil et al., Search for K^+ decays into the $\pi^+ e^+ e^- e^+ e^-$ final state. Phys. Lett. B **846**, 138193 (2023). https://doi.org/10.1016/j.physletb.2023.138193
18. G. D'Ambrosio, A. Iyer, F. Mahmoudi, S. Neshatpour, Anatomy of kaon decays and prospects for lepton flavour universality violation. J. High Energy Phys. **2209**, 148 (2022). https://doi.org/10.1007/JHEP09(2022)148
19. A. Diamant-Berger et al., Study of some rare decays of the K^+ meson. Phys. Lett. B **62**(4), 485–490 (1976). https://doi.org/10.1016/0370-2693(76)90690-0
20. V. Fanti et al., The beam and detector for the NA48 neutral kaon CP violation experiment at CERN. Nucl. Instrum. Methods Phys. Res. A **574**(3), 433–471 (2007). https://doi.org/10.1016/j.nima.2007.01.178
21. J.R. Fry, G. Ruggiero, F. Bergsma, Precision magnetic field mapping for CERN experiment NA62. J. Phys. G **43**(12), 125004 (2016). https://doi.org/10.1088/0954-3899/43/12/125004
22. E. Goudzovski et al., Development of the kaon tagging system for the NA62 experiment at CERN. Nucl. Instrum. Methods Phys. Res. A **801**, 86–94 (2015). https://doi.org/10.1016/j.nima.2015.08.015

23. P.D. Group, R.L. Workman et al., Review of Particle Physics. Prog. Theor. Exp. Phys. **2022**(8), 083C01 (2022). https://doi.org/10.1093/ptep/ptac097
24. HIKE Collaboration: High Intensity Kaon Experiments (HIKE) at the CERN SPS: Proposal for Phases 1 and 2. Tech. Rep. CERN-SPSC-2023-031, SPSC-P-368, CERN, Geneva (2023). https://cds.cern.ch/record/2878543
25. S. Kholodenko, NA62 charged particle hodoscope. Design and performance in 2016 run. J. Instrum. **12**(06), C06042 (2017). https://doi.org/10.1088/1748-0221/12/06/C06042
26. A.J. Krasznahorkay et al., Observation of anomalous internal pair creation in ^{8}Be: a possible indication of a light, neutral Boson. Phys. Rev. Lett. **116**, 042501 (2016). https://doi.org/10.1103/PhysRevLett.116.042501
27. G.A. Rinella et al., The NA62 GigaTracKer: a low mass high intensity beam 4D tracker with 65 ps time resolution on tracks. J. Instrum. **14**(07), P07010 (2019). https://doi.org/10.1088/1748-0221/14/07/P07010

Open Access This chapter is licensed under the terms of the Creative Commons Attribution 4.0 International License (http://creativecommons.org/licenses/by/4.0/), which permits use, sharing, adaptation, distribution and reproduction in any medium or format, as long as you give appropriate credit to the original author(s) and the source, provide a link to the Creative Commons license and indicate if changes were made.

The images or other third party material in this chapter are included in the chapter's Creative Commons license, unless indicated otherwise in a credit line to the material. If material is not included in the chapter's Creative Commons license and your intended use is not permitted by statutory regulation or exceeds the permitted use, you will need to obtain permission directly from the copyright holder.

Probing a Vector-like Top Quark $T \to H^+ b$ and $H^+ \to t\bar{b}$ at HL-LHC

R. Benbrik, Mbark Berrouj, and Mohammed Boukidi

1 Introduction

Vector-like quarks (VLQs) [1, 2] are heavy spin-1/2 particles predicted by various theoretical models, such as Composite Higgs models [3, 4]. Unlike Standard Model (SM) quarks, VLQs have identical Electro-Weak (EW) quantum numbers for their left- and right-handed couplings and can mix with top and bottom quarks. If their masses are within the reach of the Large Hadron Collider (LHC), VLQs could manifest through distinctive final states. The searches for VLQs at the LHC have been conducted by both the ATLAS and CMS experiments [5–7].

Previous studies [8–16] have explored the decay modes of VLQs into exotic particles ($Q \to q\Phi$ ($\Phi = H, A, H^\pm$)) within the Two Higgs Doublet Model Type II (2HDM-II). These investigations have shown potential for discovering new physics beyond the SM.

This study focuses on the single production of a vector-like top partner (T) and its decay into a charged Higgs boson ($H^\pm$) and a bottom quark. We analyze the process $pp \to qg \to T^+ b\bar{b} j \to H^+ b\bar{b} j \to W^+ b\bar{b} j \to 1\ell + 4b + 1j + \not{E}_T$, which offers promising detection potential at the HL-LHC.

2 Framework

In this section, we provide a brief overview of the 2HDM-II+TB scenario and its relevant interactions. More analytical details can be found in Ref [10].

R. Benbrik · M. Berrouj (✉) · M. Boukidi
Polydisciplinary Faculty, Laboratory of Fundamental and Applied Physics, Cadi Ayyad University, Safi, Morocco
e-mail: r.benbrik@uca.ac.ma; mbark.berrouj@ced.uca.ma; mohammed.boukidi@ced.uca.ma

The Yukawa Lagrangian for the charged Higgs interactions is given by:

$$\mathcal{L}_{H^+} = -\frac{gm_t}{\sqrt{2}M_W}\bar{t}(\cot\beta Z_{tb}^L P_L + \tan\beta Z_{tb}^R P_R)bH^+$$
$$- \frac{gm_T}{\sqrt{2}M_W}\bar{T}(\cot\beta Z_{Tb}^L P_L + \tan\beta Z_{Tb}^R P_R)bH^+ + \text{H.c.} \quad (1)$$

The interaction with the W boson is given by:

$$\mathcal{L}_W = -\frac{g}{\sqrt{2}}\bar{T}\gamma^\mu(V_{Tb}^L P_L + V_{Tb}^R P_R)bW_\mu^+ + \text{H.c.} \quad (2)$$

where the couplings are defined as:

$$Z_{tb}^L = c_L^d c_L^u + \frac{s_L^d}{s_L^u}(s_L^{u\,2} - s_R^{u\,2})e^{i(\phi_u - \phi_d)}, \quad (3)$$

$$Z_{tb}^R = \frac{m_b}{m_t}\left[c_L^u c_L^d + \frac{s_L^u}{s_L^d}(s_L^{d\,2} - s_R^{d\,2})e^{i(\phi_u - \phi_d)}\right], \quad (4)$$

$$Z_{Tb}^L = c_L^d s_L^u e^{-i\phi_u} + (s_L^{u\,2} - s_R^{u\,2})\frac{s_L^d}{c_L^u}e^{-i\phi_d}, \quad (5)$$

$$Z_{Tb}^R = \frac{m_b}{m_T}\left[c_L^d s_L^u e^{-i\phi_u} + (s_R^{d\,2} - s_L^{d\,2})\frac{c_L^u}{s_L^d}e^{-i\phi_d}\right]. \quad (6)$$

3 Numerical Analysis

We performed a numerical scan over the parameter space detailed in Ref [11] to identify scenarios with a significant cross section for the final state $W^+b\bar{b}b\bar{b}j$. The parameter points were checked for consistency with various theoretical and experimental constraints described in the same reference. The aim is to identify parameter regions within the model that are consistent with existing data and exhibit a potentially large cross section for the signal process.

For illustration, we selected three benchmark points (BPs) from the scan that show notable cross sections for varying values of m_T. The specifics of these BPs are provided in Table 1.

Table 1 The description of our BPs. Reprinted under CC-BY-4.0 from [11]. ©2023, The Author(s)

Parameters	BP_1	BP_2	BP_3
(Masses are in GeV)			
m_h	125	125	125
m_H	629.71	644.33	599.34
m_A	626.62	620.03	586.75
$m_{H^\pm}$	654.40	636.73	600.62
$\tan\beta$	3.24	2.59	2.74
$\sin(\beta-\alpha)$	1	1	1
m_T	1498.93	1591.92	1848.90
m_B	1512.33	1612.45	1870.31
$\sin(\theta^u)_L$	0.0035	0.0062	−0.0014
$\sin(\theta^d)_L$	0.0004	0.0005	0.0004
$\sin(\theta^u)_R$	0.0306	0.0575	−0.0149
$\sin(\theta^d)_R$	0.1362	0.1688	0.1516

Table 2 The set of requirements applied to select the signal events

$N(l)=1$, $\quad N(b)\geq 4$
$P_T^l > 50\,\text{GeV}$, $\quad \not{E}_T > 30\,\text{GeV}$, $\quad E_T > 1000\,\text{GeV}$
$P_T^{b_1} > 450\,\text{GeV}$ $\quad P_T^{b_2} > 220\,\text{GeV}$,
$M_T^{b_1 b_2 b_3 b_4 l} > 1450\,\text{GeV}$

4 Detector Simulation

The Monte Carlo generation of events for each signal BP was performed using a `FeynRules` [17] model file and the corresponding UFO model, based on the model description in Ref [10]. Signal and background events were generated at Leading Order (LO) using `MadGraph5_aMC_v3.3.1` [18] and processed through `Pythia-8.30` [19] for showering and hadronization. Jet clustering was performed with `FastJet` [20], and fast detector simulations were conducted using `Delphes-3.5.0` [21] with the default HL-LHC detector card, which assumes the anti-k_t algorithm with a cone radius of $R = 0.4$ Event analysis was performed using `MadAnalysis5` [22].

The primary SM background contributions arise from $t\bar{t}b\bar{b}$, $t\bar{t}jj$, and ttV ($V = Z, W, h$). Signal and background parton-level events must satisfy the following requirements:

$$|\eta(\ell)| < 2.5, \quad |\eta(j,b)| < 5, \quad p_T(x) > 25\,\text{GeV}, \quad \Delta R(x,y) > 0.4,$$

where $x, y = \ell, j, b$, $\eta(x)$ is the pseudorapidity of x, $p_T(x)$ is the transverse momentum of x, and $\Delta R(x, y)$ is the separation between two particles x and y.

A set of requirements presented in Table 2, were imposed to enhance the signal significance relative to the background.

The discovery significance, $\mathcal{Z}_{\text{disc}}$, and the exclusion significance, $\mathcal{Z}_{\text{excl}}$, are calculated using the following expressions:

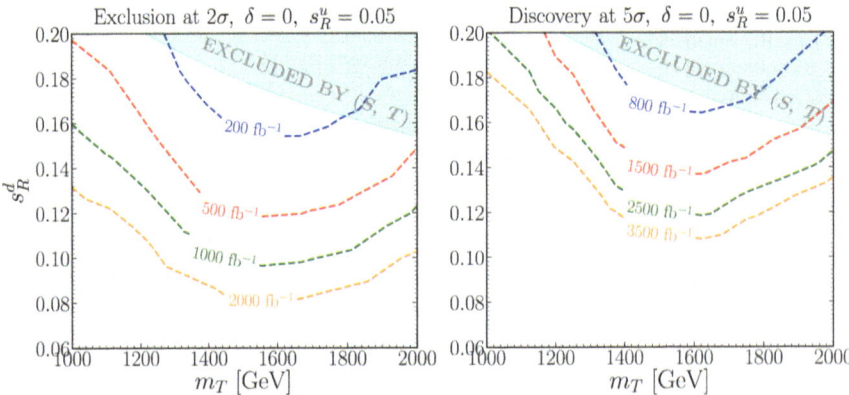

Fig. 1 The discovery prospects (at 5σ) and exclusion limit (at 2σ) contour plots for the signal in $m_T - \sin\theta_R^u$ planes at $\sqrt{s} = 14$ TeV LHC for different integrated luminosities. Reprinted under CC-BY-4.0 from [11]. ©2023, The Author(s)

$$\mathcal{Z}_{\text{disc}} = \sqrt{2\left[(s+b)\ln\left(1+s/b\right) - s\right]}, \qquad (7)$$

$$\mathcal{Z}_{\text{excl}} = \sqrt{2\left[s - b\ln\left(1+s/b\right)\right]}.$$

Here, s and b represent the expected event count of the signal and total SM background after all cuts have been applied, respectively.

Figure 1 presents the 5σ observation potential and 95% exclusion upper limit contour plots for the signal in $m_T - \sin\theta_R^u$ planes at $\sqrt{s} = 14$ TeV LHC for different integrated luminosities. With $\sin\theta_R^u = 0.05$, $\tan\beta = 2.6$, $m_H \sim m_A \sim m_{H^\pm} = 600$ GeV, and $m_{12}^2 = m_A^2 \tan\beta / \left(1 + \tan\beta^2\right)$.

5 Conclusion

The $T \to H^\pm b$ decay mode is shown to be a powerful tool for discovering new physics beyond the SM. This study demonstrates its sensitivity, particularly at high integrated luminosities.

References

1. J.A. Aguilar-Saavedra, R. Benbrik, S. Heinemeyer, M. Pérez-Victoria, Handbook of vectorlike quarks, Mixing and single production. Phys. Rev. D **88**, 094010 (2013). https://doi.org/10.1103/PhysRevD.88.094010
2. M. Buchkremer, G. Cacciapaglia, A. Deandrea, L. Panizzi, Model independent framework for searches of top partners. Nucl. Phys. B **876**, 376 (2013). https://doi.org/10.1016/j.nuclphysb.2013.08.010

3. B.A. Dobrescu, C.T. Hill, Electroweak symmetry breaking via top condensation seesaw. Phys. Rev. Lett. **81**, 2634 (1998). https://doi.org/10.1103/PhysRevLett.81.2634
4. R.S. Chivukula, B.A. Dobrescu, H. Georgi, C.T. Hill, Top quark seesaw theory of electroweak symmetry breaking. Phys. Rev. D **59**, 075003 (1999). https://doi.org/10.1103/PhysRevD.59.075003
5. ATLAS Collaboration, Search for pair-production of vector-like quarks in lepton+jets final states containing at least one b-tagged jet using the Run 2 data from the ATLAS experiment. Phys. Lett. B **854**, 138743 (2024). https://doi.org/10.1016/j.physletb.2024.138743
6. CMS Collaboration, Search for a vector-like quark $T' \to tH$ via the diphoton decay mode of the Higgs boson in proton-proton collisions at $\sqrt{s} = 13$ TeV. J. High Energy Phys. **09**, 057 (2023). https://doi.org/10.1007/JHEP09(2023)057
7. R. Benbrik, M. Boukidi, M. Ech-chaouy, S. Moretti, K. Salime, Q.-S. Yan, Vector-Like Quarks at the LHC: A Unified Perspective from ATLAS and CMS Exclusion Limits. https://arxiv.org/abs/2412.01761
8. A. Arhrib, R. Benbrik, M. Boukidi, B. Manaut, S. Moretti, Anatomy of Vector-Like Top-Quark Models in the Alignment Limit of the 2-Higgs Doublet Model Type-II. https://arxiv.org/abs/2401.16219
9. A. Arhrib, R. Benbrik, M. Boukidi, S. Moretti, Anatomy of Vector-Like Bottom-Quark Models in the Alignment Limit of the 2-Higgs Doublet Model Type-II. https://arxiv.org/abs/2403.13021
10. R. Benbrik, M. Boukidi, S. Moretti, Probing charged Higgs bosons in the two-Higgs-doublet model type II with vectorlike quarks. Phys. Rev. D **109**, 055016 (2024). https://doi.org/10.1103/PhysRevD.109.055016
11. R. Benbrik, M. Berrouj, M. Boukidi, A. Habjia, E. Ghourmin, L. Rahili, Search for single production of vector-like top partner $T \to H + b$ and $H^{\pm} \to tb$ at the LHC Run-III. Phys. Lett. B **843**, 138024 (2023). https://doi.org/10.1016/j.physletb.2023.138024
12. R. Benbrik, M. Boukidi, Phenomenology of Heavy Quark at the LHC. https://doi.org/10.5772/intechopen.1001607
13. H. Abouabid, A. Arhrib, R. Benbrik, M. Boukidi, J.E. Falaki, The oblique parameters in the 2HDM with vector-like quarks: confronting M_W CDF-II anomaly. J. Phys. G **51**, 075001 (2024). https://doi.org/10.1088/1361-6471/ad3f34
14. A. Arhrib, R. Benbrik, M. Boukidi, S. Moretti, Large Hadron Collider Signatures of Exotic Vector-Like Quarks within the 2-Higgs Doublet Model Type-II. https://arxiv.org/abs/2409.20104
15. R. Benbrik, M. Berrouj, M. Boukidi, Investigation of Charged Higgs Bosons Production from Vector-Like T Quark Decays at $e\gamma$ Collider. https://arxiv.org/abs/2408.15985
16. A. Arhrib, R. Benbrik, M. Berrouj, M. Boukidi, B. Manaut, Search for Charged Higgs Bosons through Vector-Like Top Quark Pair Production at the LHC. https://arxiv.org/abs/2407.01348
17. C. Degrande, C. Duhr, B. Fuks, D. Grellscheid, O. Mattelaer, T. Reiter, UFO - the universal FeynRules output. Comput. Phys. Commun. **183**, 1201 (2012). https://doi.org/10.1016/j.cpc.2012.01.022
18. J. Alwall, R. Frederix, S. Frixione, V. Hirschi, F. Maltoni, O. Mattelaer, et al., The automated computation of tree-level and next-to-leading order differential cross sections, and their matching to parton shower simulations. J. High Energy Phys. **07**, 079 (2014). https://doi.org/10.1007/JHEP07(2014)079
19. T. Sjöstrand, S. Ask, J.R. Christiansen, R. Corke, N. Desai, P. Ilten, et al., An introduction to PYTHIA 8.2. Comput. Phys. Commun. **191**, 159 (2015). https://doi.org/10.1016/j.cpc.2015.01.024
20. M. Cacciari, G.P. Salam, G. Soyez, FastJet user manual. Eur. Phys. J. C **72**, 1896 (2012). https://doi.org/10.1140/epjc/s10052-012-1896-2
21. DELPHES 3 Collaboration, DELPHES 3, A modular framework for fast simulation of a generic collider experiment. J. High Energy Phys. **02**, 057 (2014). https://doi.org/10.1007/JHEP02(2014)057
22. E. Conte, B. Fuks, G. Serret, MadAnalysis 5, A user-friendly framework for collider phenomenology. Comput. Phys. Commun. **184**, 222 (2013). https://doi.org/10.1016/j.cpc.2012.09.009

Open Access This chapter is licensed under the terms of the Creative Commons Attribution 4.0 International License (http://creativecommons.org/licenses/by/4.0/), which permits use, sharing, adaptation, distribution and reproduction in any medium or format, as long as you give appropriate credit to the original author(s) and the source, provide a link to the Creative Commons license and indicate if changes were made.

The images or other third party material in this chapter are included in the chapter's Creative Commons license, unless indicated otherwise in a credit line to the material. If material is not included in the chapter's Creative Commons license and your intended use is not permitted by statutory regulation or exceeds the permitted use, you will need to obtain permission directly from the copyright holder.

Interpreting the Indications of a 95 GeV Higgs Boson Through a 2-Higgs Doublet Model

A. Belyaev, R. Benbrik, Mohammed Boukidi, M. Chakraborti, S. Moretti, and S. Semlali

1 Introduction

Since the discovery of the Higgs boson at the Large Hadron Collider (LHC), extensive efforts have focused on accurately determining its characteristics. While the majority of findings are consistent with the predictions of the Standard Model (SM), there has been significant interest in exploring the existence of additional Higgs bosons. Notably, the CMS [16] collaboration reported an excess with a local (global) significance of 2.9σ (1.3σ) in the di-photon channel around 95 GeV, utilizing data from the LHC at $\sqrt{s} = 13$ TeV. Concurrently, the ATLAS collaboration identified

---

The author "Mohammed Boukidi" is speaker of this chapter.

A. Belyaev · S. Semlali
Particle Physics Department, Rutherford Appleton Laboratory, Chilton, UK

School of Physics and Astronomy, University of Southampton, Southampton, UK
e-mail: a.belyaev@soton.ac.uk; souad.semlali@soton.ac.uk

R. Benbrik · M. Boukidi (✉)
Polydisciplinary Faculty, Laboratory of Fundamental and Applied Physics, Cadi Ayyad University, Sidi Bouzid, Safi, Morocco
e-mail: r.benbrik@uca.ac.ma; mohammed.boukidi@ced.uca.ma

M. Chakraborti
School of Physics and Astronomy, University of Southampton, Southampton, UK
e-mail: mani.chakraborti@gmail.com

S. Moretti
School of Physics and Astronomy, University of Southampton, Southampton, UK

Department of Physics and Astronomy, Uppsala University, Uppsala, Sweden
e-mail: s.moretti@soton.ac.uk; stefano.moretti@physics.uu.se

a similar excess with a local significance of 1.7σ [1]. Additional discrepancies in the $\tau\tau$ [14] and $b\bar{b}$ [4] channels have been observed by the CMS and LEP collaborations, respectively, prompting investigations into Beyond the SM (BSM) explanations. Among various models, the 2-Higgs Doublet Model (2HDM) [11, 19], particularly its Type-III representation [7–9], stands out as a viable framework to account for these anomalies. This paper focuses on the 2HDM Type-III's capacity to simultaneously explain the observed excesses in the $\gamma\gamma$, $\tau\tau$, and $b\bar{b}$ channels, alongside its predictions for modifications in the Higgs boson production rate, especially in the $gg, qq \to h_{125}t\bar{t}$ process.

2 2HDM Type-III

The 2HDM includes two $SU(2)_L$ doublets with hypercharge $Y = 1$. The most general renormalisable $SU(2)_L \times U(1)_Y$ invariant scalar potential is written as follows [11]:

$$\mathcal{V} = m_{11}^2 \Phi_1^\dagger \Phi_1 + m_{22}^2 \Phi_2^\dagger \Phi_2 - \left[m_{12}^2 \Phi_1^\dagger \Phi_2 + \text{H.c.} \right] + \lambda_1 (\Phi_1^\dagger \Phi_1)^2 + \lambda_2 (\Phi_2^\dagger \Phi_2)^2$$

$$+ \lambda_3 (\Phi_1^\dagger \Phi_1)(\Phi_2^\dagger \Phi_2) \lambda_4 (\Phi_1^\dagger \Phi_2)(\Phi_2^\dagger \Phi_1) + \tfrac{1}{2} \left[\lambda_5 (\Phi_1^\dagger \Phi_2)^2 + \text{H.c.} \right]$$

$$+ \left\{ \left[\lambda_6 (\Phi_1^\dagger \Phi_1) + \lambda_7 (\Phi_2^\dagger \Phi_2) \right] (\Phi_1^\dagger \Phi_2) + \text{H.c.} \right\} \tag{1}$$

Adopting $\mathcal{CP}$-conserving option, and very minimal version of Higgs couplings, the above potential can be parametrized by seven free parameters, those are: Higgs bosons masses, $m_h, m_H, m_{H^\pm}, m_A$, the ratio of the vacuum expectation values of the two Higgs doublets fields $\tan \beta = v_2/v_1$, the mixing angle of the $\mathcal{CP}$-even Higgs states α, and m_{12}^2. In the Yukawa sector, the general scalar to fermions couplings are given by:

$$-\mathcal{L}_Y = \bar{Q}_L Y_1^u U_R \tilde{\Phi}_1 + \bar{Q}_L Y_2^u U_R \tilde{\Phi}_2 + \bar{Q}_L Y_1^d D_R \Phi_1 + \bar{Q}_L Y_2^d D_R \Phi_2$$

$$+ \bar{L} Y_1^\ell \ell_R \Phi_1 + \bar{L} Y_2^\ell \ell_R \Phi_2 + \text{H.c.}, \tag{2}$$

where $Y_{1,2}^f$ are 3×3 Yukawa matrices in flavour space and $\tilde{\Phi}_{1,2} = i\sigma_{1,2}\Phi_{1,2}^*$, with $\sigma_{1,2}$ being the Pauli matrices. After EWSB has taken place, one can then derive the fermion masses from Eq. (2).

Here, however, we investigate a modified version of the described 2HDM, the so-called Type-III, where neither a global symmetry is implemented in the Yukawa sector nor any alignment in flavour space is enforced. We adopt instead the Cheng-Sher ansatz [13, 17], which assumes a flavour symmetry in turn suggesting a specific texture of the Yukawa matrices, where FCNC effects are proportional to the geometric mean of the two fermion masses and dimensionless parameters[1] χ_{ij}^f ($\propto \sqrt{m_i m_j}/v \, \chi_{ij}^f$), where $i, j = 1 - 3$. After EWSB, the Yukawa sector can be expressed in terms of the mass eigenstates of the Higgs bosons, as follows:

$$-\mathcal{L}_Y^{III} = \sum_{f=u,d,\ell} \frac{m_j^f}{v} \times \left((\xi_h^f)_{ij} \bar{f}_{Li} f_{Rj} h + (\xi_H^f)_{ij} \bar{f}_{Li} f_{Rj} H - i(\xi_A^f)_{ij} \bar{f}_{Li} f_{Rj} A \right)$$

$$+ \frac{\sqrt{2}}{v} \sum_{k=1}^{3} \bar{u}_i \left[\left(m_i^u (\xi_A^{u*})_{ki} V_{kj} P_L + V_{ik} (\xi_A^d)_{kj} m_j^d P_R \right) \right] d_j H^+$$

$$+ \frac{\sqrt{2}}{v} \bar{\nu}_i (\xi_A^\ell)_{ij} m_j^\ell P_R \ell_j H^+ + \text{H.c.}, \tag{3}$$

where the $(\xi_\phi^f)_{ij}$ couplings are given in Ref. [7] in terms of the free parameters χ_{ij}^f, $\tan \beta$ and the mixing angle α.

3 The Excesses in the $h \to \gamma\gamma$, $\tau\tau$, and $b\bar{b}$ Channels

We explore the compatibility of the 2HDM Type-III with the observed excesses at 94–97 GeV in the $\gamma\gamma$, $\tau\tau$, and $b\bar{b}$ channels as reported by both LEP and the LHC. The analysis is performed under the Narrow Width Approximation (NWA), focusing on signal strengths derived from the product of cross sections (σ) and decay Branching Ratios ($\mathcal{BR}$s), specifically:

$$\mu_{b\bar{b}} = \frac{\sigma_{2\text{HDM}}(e^+e^- \to Z\phi)}{\sigma_{\text{SM}}(e^+e^- \to Zh_{\text{SM}})} \times \frac{\mathcal{BR}_{2\text{HDM}}(\phi \to b\bar{b})}{\mathcal{BR}_{\text{SM}}(h_{\text{SM}} \to b\bar{b})}, \tag{4}$$

$$\mu_{\gamma\gamma,\tau\tau} = \frac{\sigma_{2\text{HDM}}(gg \to \phi)}{\sigma_{\text{SM}}(gg \to h_{\text{SM}})} \times \frac{\mathcal{BR}_{2\text{HDM}}(\phi \to \gamma\gamma, \tau\tau)}{\mathcal{BR}_{\text{SM}}(h_{\text{SM}} \to \gamma\gamma, \tau\tau)}. \tag{5}$$

where ϕ represents the light CP-even Higgs scalar h, and $c_{\phi ZZ}$ and $c_{\phi tt}$ denote its couplings to ZZ and $t\bar{t}$, normalized to the SM values.

Experimental measurements for these channels are:

$$\mu_{\gamma\gamma}^{\exp(\text{ATLAS+CMS})} = 0.24^{+0.09}_{-0.08}, \; \mu_{\tau\tau}^{\exp} = 1.2 \pm 0.5, \; \mu_{b\bar{b}}^{\exp} = 0.117 \pm 0.057. \tag{6}$$

with the combined di-photon signal strength from ATLAS and CMS measurements indicating a 3.1σ local excess, as computed by Ref. [10]. A χ^2 analysis is conducted using measured central values $\mu^{\exp}$ and the 1σ uncertainties $\Delta\mu^{\exp}$ of the signal

---

[1] For further details, see Ref. [20].

rates related to the two excesses as defined in Eqs. (4)–(5). The contribution to the χ^2 value for each channel ($\gamma\gamma$, $\tau\tau$, $b\bar{b}$) is calculated using the formula

$$\chi^2_{\gamma\gamma,\tau\tau,b\bar{b}} = \frac{(\mu_{\gamma\gamma,\tau\tau,b\bar{b}} - \mu^{\text{exp}}_{\gamma\gamma,\tau\tau,b\bar{b}})^2}{(\Delta\mu^{\text{exp}}_{\gamma\gamma,\tau\tau,b\bar{b}})^2}. \tag{7}$$

So, the resulting χ^2 which we will use to judge whether the points from the model describe the excess in three channels, reads as:

$$\chi^2_{95=\gamma\gamma+\tau\tau+b\bar{b}} = \chi^2_{\gamma\gamma} + \chi^2_{\tau\tau} + \chi^2_{b\bar{b}}. \tag{8}$$

4 Numerical Analysis

Before presenting our results, we describe the theoretical self-consistency requirements and experimental measurements used to constrain the parameter space of the 2HDM Type-III scenario. The theoretical requirements include the perturbativity of scalar quartic couplings, vacuum stability, and tree-level perturbative unitarity conditions for various scattering amplitudes of gauge and Higgs boson states. These constraints were tested using the public code 2HDMC-1.8.0 [18]. On the experimental side, we considered constraints from electroweak precision observables (EWPO) in terms of the oblique parameters S, T, and U. We also took into account measurements of the newly discovered Higgs boson properties at the LHC, cross-section limits from LEP, Tevatron, and the LHC for additional Higgs boson searches using HiggsBounds-6 [6], and current Higgs boson signal strength measurements using HiggsSignals-3 [5]. These packages are part of HiggsTools [2]. Additionally, we included constraints from flavour physics using SuperIso-4.1 [21]. Subsequently, we performed a systematic random scan over the specified parameter ranges (Table 1).

In Fig. 1, we show $\chi^2_{\gamma\gamma+\tau\tau+b\bar{b}}$ results as a color map projected onto the ($\mu_{\tau\tau}$-$\mu_{\gamma\gamma}$) (left), ($\mu_{b\bar{b}}$-$\mu_{\gamma\gamma}$) (middle), and ($\mu_{\tau\tau}$-$\mu_{b\bar{b}}$) (right) planes of the signal strength parameters. The dashed ellipses indicate regions consistent with the excess at 1σ. Black, gray, and red contours represent χ^2 values using $\mu^{\text{CMS}}_{\gamma\gamma}$, $\mu^{\text{ATLAS}}_{\gamma\gamma}$, and $\mu^{\text{CMS+ATLAS}}_{\gamma\gamma}$, respectively. The value of $\chi^2_{\gamma\gamma+\tau\tau+b\bar{b}}$ is indicated by the color map. Grey points show exclusions based on recent CMS searches. The green star marks

Table 1 Scan ranges of the 2HDM Type-III input parameters. Masses are in GeV

m_h	m_H	m_A	$m_{H^\pm}$	$s_{\beta-\alpha}$	$\tan\beta$	m^2_{12}	$\chi^{f,\ell}_{ij}$
[94; 97]	125.09	[80; 300]	[160; 200]	[−0.5; 0]	[1; 30]	$m^2_h \tan\beta/(1+\tan^2\beta)$	[−3; 3]

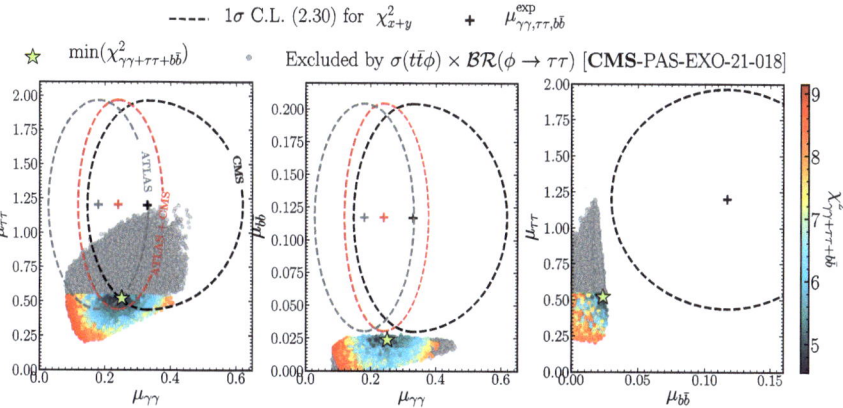

Fig. 1 The colour map of $\chi^2_{\gamma\gamma+\tau\tau+b\bar{b}}$ in the $(\mu_{\gamma\gamma} - \mu_{\tau\tau})$, $(\mu_{\gamma\gamma} - \mu_{b\bar{b}})$ and $(\mu_{b\bar{b}} - \mu_{\tau\tau})$ planes of the signal strength parameters for 2HDM Type-III parameter space under study. The ellipses define the regions consistent with the excess at 1σ C.L. The contribution from $\mu_{\gamma\gamma}$ to χ^2 for black, gray and red contours comes from CMS, ATLAS and combined CMS+ATLAS data respectively. Grey points are excluded by $\sigma(t\bar{t}\phi) \times \mathcal{BR}(\phi \to \tau\tau)$. The position of χ^2_{95min} is marked by a green star. Reprinted under CC-BY-4.0 license from [15]. © 2022, CERN for the benefit of the CMS Collaboration

the minimum of $\chi^2_{\gamma\gamma+\tau\tau+b\bar{b}}$ at 4.55, corresponding to a 1.26 σ C.L. for three degrees of freedom, within the 1σ C.L. contour for $\mu_{\tau\tau}$-$\mu_{\gamma\gamma}$ but not for the other pairs. Dark blue points around this minimum indicate the 2HDM Type-III can explain the excess at 1.5σ C.L or better, reaching up to 1.3σ. The middle and right frames show that all points are outside the 1σ ellipses, as it is difficult to achieve large $\mu_{b\bar{b}}$ values while satisfying all constraints. This is due to the $\gamma\gamma$ excess being achieved by enhancing $\mathcal{BR}(h_{95} \to \gamma\gamma)$ through a decrease in $\Gamma_{b\bar{b}}$, the main decay channel, thus reducing $\mathcal{BR}(h_{95} \to b\bar{b})$.

Figure 2 illustrates the correlation between the normalized couplings $|c_{h_{125}VV}|$ and $|c_{h_{125}t\bar{t}}|$ of the ≈ 125 GeV Higgs. Explaining the excesses in $b\bar{b}$, $\tau\tau$, and $\gamma\gamma$ channels requires an enhancement of the $h_{125}t\bar{t}$ and $h_{125}b\bar{b}$ couplings by about 11.5% from the SM. The blue and orange ellipses represent the projected experimental precision for the normalized couplings at the HL-LHC with 3000 fb^{-1} and the ILC 500, respectively. The centers of these projections, corresponding to the SM value, are represented by black diamond. Points describing the excesses lie outside these ellipses, indicating that the precision of these experiments would allow distinguishing between the SM-like properties of h_{125} and the 2HDM Type-III model. Thus, the HL-LHC and ILC 500 could either confirm or rule out the scenario explaining the current excesses in the three channels.

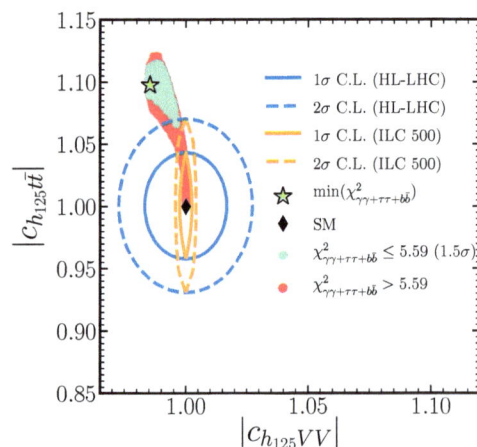

Fig. 2 Correlation between normalized couplings $|c_{h_{125}VV}|$ and $|c_{h_{125}t\bar{t}}|$. The blue and orange ellipses indicate the projected uncertainties at the HL-LHC, and following a combination of data from both the HL-LHC and the ILC at a center-of-mass energy of 500 GeV, respectively. Reprinted under CC-BY-4.0 license from [3, 12]. © 2019, CERN & © 2019, The Author(s)

5 Conclusion

In this contribution, we have shown that within the 2HDM Type-III, the observed excesses in the di-photon, di-tau, and $b\bar{b}$ channels can be attributed to the lightest CP-even Higgs state, h, with a mass around 95 GeV. This finding aligns with both theoretical requirements and current experimental constraints. Our analysis indicates that these excesses can be simultaneously explained at approximately 1.3σ C.L., requiring an enhancement in the $h_{125}t\bar{t}$ coupling by up to 12% from the SM predictions.

The emergence of a 95 GeV resonance and the associated enhancements in Yukawa couplings to top quarks make this scenario particularly compelling for further experimental investigations at the LHC, HL-LHC, and future colliders.

Acknowledgments AB and SM are supported in part through the NExT Institute and STFC CG ST/L000296/1. AB would like to thank Prof. Glen Cowan for discussions around the statistical aspects of our study. The work of RB and MB is supported by the Moroccan Ministry of Higher Education and Scientific Research MESRSFC and CNRST Project PPR/2015/6. MB acknowledges the use of CNRST/HPC-MARWAN in completing this work. SS is supported in full by the NExT Institute.

References

1. ATLAS, Search for resonances in the 65 to 110 GeV diphoton invariant mass range using 80 fb^{-1} of pp collisions collected at $\sqrt{s} = 13$ TeV with the ATLAS detector. ATLAS-CONF-2018-025 (2018)
2. Bahl, H., Biekötter, T., Heinemeyer, S., Li, C., Paasch, S., Weiglein, G., Wittbrodt, J.: HiggsTools: BSM scalar phenomenology with new versions of HiggsBounds and HiggsSignals. Comput. Phys. Commun. **291**, 108803 (2023). https://doi.org/10.1016/j.cpc.2023.108803

3. Bambade, P., et al.: The International Linear Collider: A Global Project (2019). arXiv:1903.01629
4. Barate, R., et al.: Search for the standard model Higgs boson at LEP. Phys. Lett. B **565**, 61–75 (2003). https://doi.org/10.1016/S0370-2693(03)00614-2
5. Bechtle, P., Dercks, D., Heinemeyer, S., Klingl, T., Stefaniak, T., Weiglein, G., Wittbrodt, J.: HiggsBounds-5: Testing Higgs Sectors in the LHC 13 TeV Era. Eur. Phys. J. C **80**(12), 1211 (2020). https://doi.org/10.1140/epjc/s10052-020-08557-9
6. Bechtle, P., Heinemeyer, S., Stal, O., Stefaniak, T., Weiglein, G.: Applying Exclusion Likelihoods from LHC Searches to Extended Higgs Sectors. Eur. Phys. J. C **75**(9), 421 (2015). https://doi.org/10.1140/epjc/s10052-015-3650-z
7. Belyaev, A., Benbrik, R., Boukidi, M., Chakraborti, M., Moretti, S., Semlali, S.: Explanation of the hints for a 95 GeV Higgs boson within a 2-Higgs Doublet Model. J. High Energy Phys. **05**, 209 (2024). https://doi.org/10.1007/JHEP05(2024)209
8. Benbrik, R., Boukidi, M., Moretti, S., Semlali, S.: Explaining the 96 GeV Di-photon anomaly in a generic 2HDM Type-III. Phys. Lett. B **832**, 137245 (2022). https://doi.org/10.1016/j.physletb.2022.137245
9. Benbrik, R., Boukidi, M., Moretti, S.: Superposition of CP-even and CP-odd Higgs resonances: Explaining the 95 GeV excesses within a two-Higgs-doublet mode. Phys. Rev. D. **110**(11), 115030 (2024). https://doi.org/10.1103/PhysRevD.110.115030
10. Biekötter, T., Heinemeyer, S., Weiglein, G.: 95.4 GeV diphoton excess at ATLAS and CMS. Phys. Rev. D. **109**(3), 035005 (2024). https://doi.org/10.1103/PhysRevD.109.035005
11. Branco, G.C., Ferreira, P.M., Lavoura, L., Rebelo, M.N., Sher, M., Silva, J.P.: Theory and phenomenology of two-Higgs-doublet models. Phys. Rept. **516**, 1–102 (2012). https://doi.org/10.1016/j.physrep.2012.02.002
12. Cepeda, M., et al.: Report from Working Group 2: Higgs Physics at the HL-LHC and HE-LHC. CERN Yellow Rep. Monogr. **7**, 221–584 (2019). https://doi.org/10.23731/CYRM-2019-007.221
13. Cheng, T.P., Sher, M.: Mass Matrix Ansatz and Flavor Nonconservation in Models with Multiple Higgs Doublets. Phys. Rev. D **35**, 3484 (1987). https://doi.org/10.1103/PhysRevD.35.3484
14. CMS, Searches for additional Higgs bosons and vector leptoquarks in $\tau\tau$ final states in proton-proton collisions at $\sqrt{s} = 13$ TeV (2022)
15. CMS, Search for dilepton resonances from decays of (pseudo)scalar bosons produced in association with a massive vector boson or top quark anti-top quark pair at $\sqrt{s} = 13$ TeV. CMS-PAS-EXO-21-018, CERN, Geneva, Switzerland (2022)
16. CMS, Search for a standard model-like Higgs boson in the mass range between 70 and 110 GeV in the diphoton final state in proton-proton collisions at $\sqrt{s} = 13$ TeV. CMS-PAS-HIG-20-002 (2023). https://doi.org/https://cds.cern.ch/record/2852907/files/HIG-20-002-pas.pdf
17. Diaz-Cruz, J.L., Noriega-Papaqui, R., Rosado, A.: Mass matrix ansatz and lepton flavor violation in the THDM-III. Phys. Rev. D **69**, 095002 (2004). https://doi.org/10.1103/PhysRevD.69.095002
18. Eriksson, D., Rathsman, J., Stal, O.: 2HDMC: Two-Higgs-Doublet Model Calculator Physics and Manual. Comput. Phys. Commun. **181**, 189–205 (2010). https://doi.org/10.1016/j.cpc.2009.09.011
19. Gunion, J.F., Haber, H.E., Kane, G.L., Dawson, S.: Errata for the Higgs hunter's guide (1992). arXiv:SCIPP-92-58
20. Hernandez-Sanchez, J., Moretti, S., Noriega-Papaqui, R., Rosado, A.: Off-diagonal terms in Yukawa textures of the Type-III 2-Higgs doublet model and light charged Higgs boson phenomenology. J. High Energy Phys. **07**, 044 (2013). https://doi.org/10.1007/JHEP07(2013)044
21. Mahmoudi, F.: SuperIso v2.3: A Program for calculating flavor physics observables in Supersymmetry. Comput. Phys. Commun. **180**, 1579–1613 (2009). https://doi.org/10.1016/j.cpc.2009.02.017

Open Access This chapter is licensed under the terms of the Creative Commons Attribution 4.0 International License (http://creativecommons.org/licenses/by/4.0/), which permits use, sharing, adaptation, distribution and reproduction in any medium or format, as long as you give appropriate credit to the original author(s) and the source, provide a link to the Creative Commons license and indicate if changes were made.

The images or other third party material in this chapter are included in the chapter's Creative Commons license, unless indicated otherwise in a credit line to the material. If material is not included in the chapter's Creative Commons license and your intended use is not permitted by statutory regulation or exceeds the permitted use, you will need to obtain permission directly from the copyright holder.

Search for New Phenomena with the ATLAS Detector at the LHC

Sijing Zhang

1 Introduction

The standard model (SM) of particle physics has demonstrated many triumphs along the years since its formulation, notably with the discovery of the Higgs boson by the ATLAS and CMS collaborations in 2012. However, we know that the SM is not complete and cannot describe all phenomena we observe in nature, such as the origin of dark matter and neutrino masses, the fine-tuning of the Higgs Boson mass, or the observed pattern of masses and mixing angles in the quark and lepton sectors. Numerous models have been made to extend the SM in order to solve some of the shortcomings. There is a large variety of theories beyond the standard model (BSM) that address this issues and that have a rich phenomenology leading to observable detector signatures.

One of the key focuses of the ATLAS experiment [1] is the search for new BSM phenomena. The ATLAS experiment has collected 139 fb^1 of data during the LHC Run 2 operation period in the years of 2015 to 2018. This unprecedented amount of data makes it possible to extend the reach of physics searches and

---

Sijing Zhang on behalf of the ATLAS Collaboration.

© Copyright 2024 CERN for the benefit of the ATLAS Collaboration. Reproduction of this article or parts of it is allowed as specified in the CC-BY-4.0 license.

S. Zhang (✉)
Laboratoire des 2 Infinis, Toulouse, France

Southern Methodist University, Dallas, TX, USA
e-mail: sijing.zhang@cern.ch

to develop dedicated analyses to explore for unconventional signatures. In the following sections, recent results of BSM searches are presented using the Run2 datasets recorded by the ATLAS detector.

2 Electroweak Symmetry Breaking (EWSB)

2.1 Natural EWSB: Supersymmetry (SUSY)

For the naturalness in the electroweak hierarchy, we prefer a model where we could compute where the Higgs scale comes from, without putting in tiny or huge numbers "by hand" as in the SM, e.g. SUSY, composite Higgs and neutral naturalness. In terms of SUSY, it predicts the existence of a bosonic (fermionic) partner for each fermionic (bosonic) particle of the SM. All the superpartners of the top and bottom quarks (i.e. the stop, $\tilde{t}$, and sbottom, $\tilde{b}$), the gluon (i.e. the gluino, $\tilde{g}$) and of an extended Higgs sector (i.e. the higgsinos, $\tilde{H}$) are required to be light by the class of "natural" SUSY models. Theoretically, such particles can be abundantly produced in the proton-proton (pp) collisions of the Large Hadron Collider (LHC). Exclusion limits on the higgsino masses have been set by the ATLAS and CMS experiments, which brings up the change that the higgsino could potentially be the first SUSY particle to be detected at the LHC.

The results searching for pair production of higgsinos in events with two Higgs bosons and missing transverse momentum will be presented in this section [2]. It targets the decay of the two Higgs bosons into $b\bar{b}$ pairs, leading to a reconstructed final state with at least three energetic b-jets and missing transverse momentum (E_T^{miss}). Two complementary analysis channels are used, with each of them targeting either low- or high-mass range of the higgsino mass. Exclusion limits as a function of the higgsino decay branching ratio to a Z or a Higgs boson are obtained. As shown in Fig. 1, masses up to 940 GeV are excluded for higgsinos decaying exclusively to Higgs bosons at 95% confidence level (CL). No significant excess above the Standard Model prediction is found.

The SUSY partners of the Higgs bosons and the SM electroweak gauge bosons, collectively called electroweakinos, are the higgsinos, winos (partners of the $SU(2)_L$ gauge fields), and bino (partner of the U(1) gauge field). The direct production of electroweakinos analysis [3] looks for chargino pairs, $\tilde{\chi}_1^+ \tilde{\chi}_1^-$, chargino and next-to-lightest neutralino, $\tilde{\chi}_1^\pm \tilde{\chi}_2^0$. Model-dependent exclusion limits at 95% CL are set. Chargino masses ranging from 260 to 520 GeV are excluded for a massless $\tilde{\chi}_1^0$ in chargino pair production models. Degenerate chargino and next-to-lightest neutralino masses ranging from 260 to 420 GeV are excluded for a massless $\tilde{\chi}_1^0$ for $\tilde{\chi}_2^0 \to Z\tilde{\chi}_1^0$. No deviations from the Standard Model expectations were found.

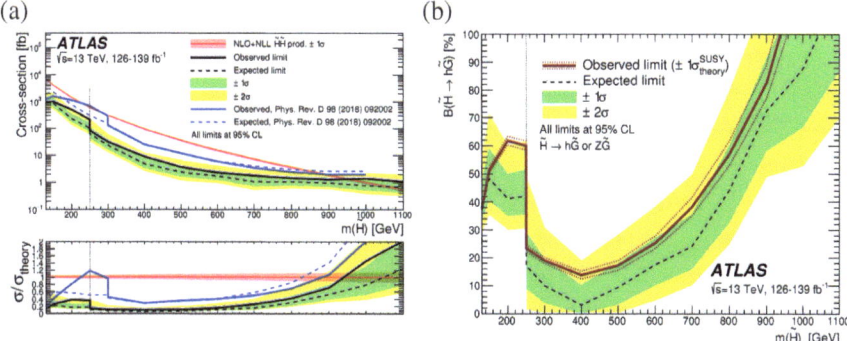

Fig. 1 The observed (solid) and expected (dashed) 95% CL upper limits on the cross section of higgsino pair production (**a**) and on BR($\tilde{H} \to h + \tilde{G}$) (**b**). The results of both low mass channel ($m_{\tilde{H}}<250$ GeV) and high mass channel ($m_{\tilde{H}}>250$ GeV) are included. The phase space above the lines are excluded in both the two plots. Reprinted under CC-BY-4.0 from [2]. © 2024, CERN for the ATLAS Collaboration

2.2 Natural EWSB: Large/Warped Extra Dimensions

Low-scale quantum gravity models introduce the existence of Quantum Black Holes (QBHs) that could provide solutions to the mass hierarchy problem of the Standard Model (SM). In low-scale quantum gravity models, this is achieved by means of lowering the scale of quantum gravity (M_D) from the high Planck-scale ($\sim 10^{16}$ TeV) to the TeV region (1–10 TeV). The search for QBHs can be conducted by examining the lepton+jets invariant mass spectra. The latest results at $\sqrt{s}=$ 13 TeV [4] are compatible with the Standard Model expectations. No excess was observed in the distribution of the invariant mass spectrum of lepton+jet pairs. Upper limit for the combined lepton+jet channel for the production of a QBH in the Arkani-Hamed-Dimopoulos-Dvali (ADD) and Randall-Sundrum (RS1) models on the cross-section times branching fraction ($\sigma \times Br$) as a function of M_{th} (threshold mass of QBHs) is shown in Fig. 2. No excess is observed for a QBH signal at any M_{th} in both models. The observation shows that invariant mass spectrum of lepton+jet pairs is consistent with SM expectations.

3 Neutrino Mass/Mixing, Flavor Anomalies Studies

Many BSM models predict the existence of new, massive particles that decay into hadronically interacting particles. The search for a massive resonance in the final states of pairs of dijet resonances is performed with the 139 fb^{-1} Run 2 datasets collected by the ATLAS detector [5]. The resonances are searched in both the invariant mass of the tetrajet system and the average invariant mass of the dijet pair systems. The background was estimated using the data-driven method to fit the

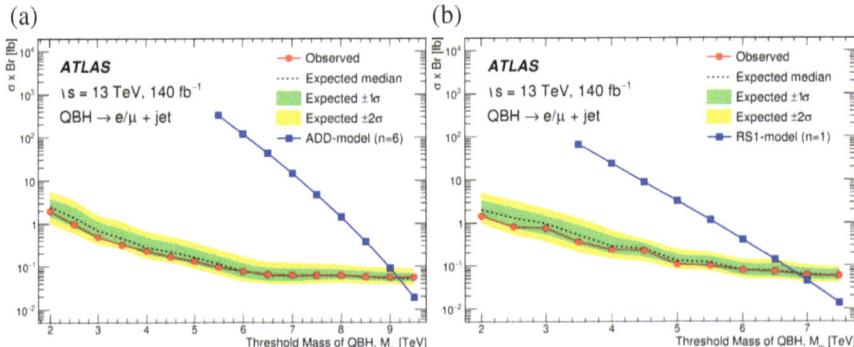

Fig. 2 The combined 95% CL upper limits on $\sigma \times Br$ as a function of M_{th} for QBH production at $M_{th} = M_D$ with decay into lepton+jet for (**a**) ADD (extra dimensions $n = 6$) and (**b**) RS1 (extra dimensions $n = 1$). Reprinted under CC-BY-4.0 license from [4]. © 2024, CERN for the ATLAS Collaboration

tetrajet and dijet invariant mass distributions with a four-parameter dijet function. There is no significant excess of events beyond the Standard Model expectation. Upper limits are set on the production cross-sections of new physics scenarios.

The results can be interpreted using the R-parity-violating supersymmetry models, in which feature prompt gluino-pair production that decays either directly to three jets, or to two jets and a neutralino, which eventually decays promptly to three jets. No significant excess over the Standard Model expectation is observed. Exclusion limits at the 95% CL are obtained. As shown in Fig. 3, the gluino mass range higher than 1800 GeV are excluded in the direct decay mode with three jets final states. For the cascade case, gluino mass range less than 2340 GeV are excluded for a neutralino with mass up to 1250 GeV.

4 Dynamic Explanation of EWSB Studies

New gauge bosons like W' or Z' bosons are predicted by multiple theories beyond the Standard Model (SM). The vector boson resonances search for the new W' is performed in the final state of a top quark and a bottom quark with the 139 fb^{-1} dataset collected by the ATLAS detector [7]. The reconstructed tb invariant mass is used for the goal of this analysis and the scenarios of purely right-handed or left-handed chirality are considered to interpreted the results in a mass range of 0.5–6 TeV. In addition, the coupling of the W' boson to the top and bottom quarks with different values are considered, including the interference with single-top-quark production in the s-channel. No significant excess from the background prediction is observed. Figure 4 shows the upper limits on the $W' \to tb$ production cross-section times branching ratio as a function of the W'-boson mass and in the plane of the coupling versus the W'-boson mass.

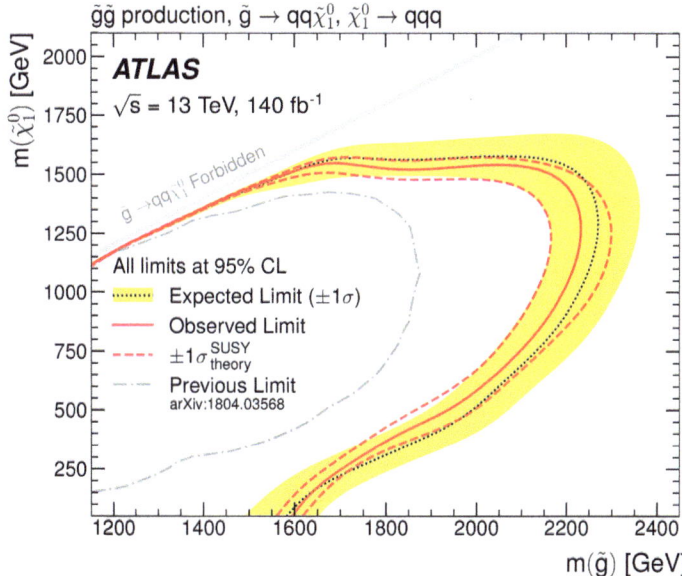

Fig. 3 Observed and expected 95% CL exclusion contours for the gluino cascade decay model with UDB decays. Reprinted under CC-BY-4.0 license from [6].© 2024, CERN for the benefit of the ATLAS Collaboration

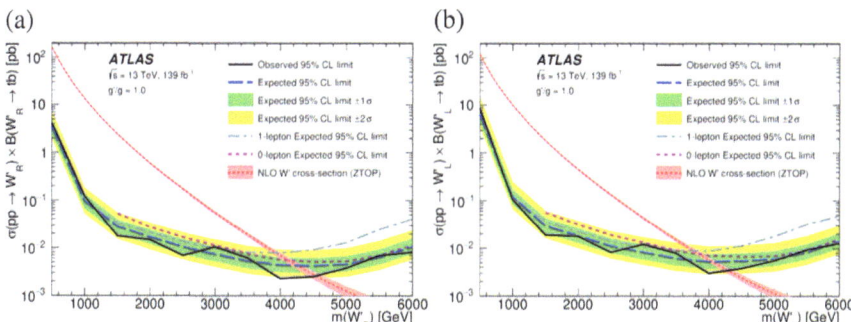

Fig. 4 Observed and expected 95% CL limits on the cross-section times branching ratio for the production of a W' boson with decay into tb, right-handed (**a**) and left-handed (**b**) couplings as a function of the mass of the W' boson and a coupling value of $g'/g = 1.0$. Reprinted under CC-BY-4.0 license from [7].© 2024, CERN for the benefit of the ATLAS Collaboration

Two vector-like quarks searches are performed with the ATLAS Run 2 datasets, one of which searches for pair-production of heavy vector-like quarks (T) that decay into a W boson and a b quark (one W boson decays leptonically [8] while the other decays hadronically) [9]. In the leptonic decay search [8], vector-like T quarks with 100% branching ratio to Wb are excluded for the mass range $m_T < 1700$ GeV. Isospin singlets with Branching ratio ($T \to Wb : Ht : Zt$) =

1/2 : 1/4 : 1/4 are excluded for the mass range m_T <1420 GeV. The hadronic decay search [9] looks for single production of a vector-like B quark in the final state of a Standard Model b-quark and a Standard Model Higgs boson (decays into a $b\bar{b}$ pair) with the ATLAS Run 2 datasets. No significant excess over the Standard Model background prediction is observed. Exclusion limits with mass-dependence at the 95% confidence level are set based on the resonance production cross-section in several theoretical scenarios.

5 Unification of All Forces Studies

The Grand Unified Theories (GUTs) [10] predict that the strong, weak and electromagnetic forces would merge into a single force at extremely high energy levels (around 10^{15} to 10^{16} GeV). Many GUT models incorporate SUSY, which proposes a symmetry between fermions and bosons, as well as the magnetic monopoles. Searches for magnetic monopoles and high-electric-charge objects (HECOs) have been extended from cosmic rays to colliders. Magnetic monopoles or HECOs would leave large energy deposits along their trajectories in the ATLAS detector. Since the ATLAS detector was designed to record low-charge and neutral particles, the characterisation of these high-energy deposits is vital to the search. A recent search for magnetic monopoles and HECOs has been performed using the ATLAS full Run 2 datasets [11]. In this search candidate signal events are recorded using a dedicated high ionizing particle (HIP) trigger, and reconstructed using the deposits in the Transition Radiation Tracker and the electromagnetic calorimeter. The benchmark models considered in this analysis are the Drell-Yan and photon-fusion pair production mechanisms. As the first search on the photon-fusion pair production mechanism of magnetic monopoles and HECOs performed with the ATLAS detector, the upper limits on the cross-section were set for spin-0 and spin-1/2 magnetic monopoles for the HECOs of electric charge $20 \leq |z| \leq 100$, as well as the mass range of 200 GeV $< m <$ 4000 GeV. Figure 5 shows the observed 95% CL upper limits as a function of the HIP mass and the comparison of the lower mass limits observed by the LHC Run 1 and Run 2 searches. There is no evidence for the existence of HIP candidate.

6 Dark Matter

As an important topic of the new phenomena, several analyses including searching for the axion-like particles, dark sector, SUSY and scalar mediators have performed with the ATLAS Run 2 datasets. The search which looks for new phenomena with top quark pairs carries out in final states with one isolated lepton, either an electron or a muon, jets, and large missing transverse momentum. In the

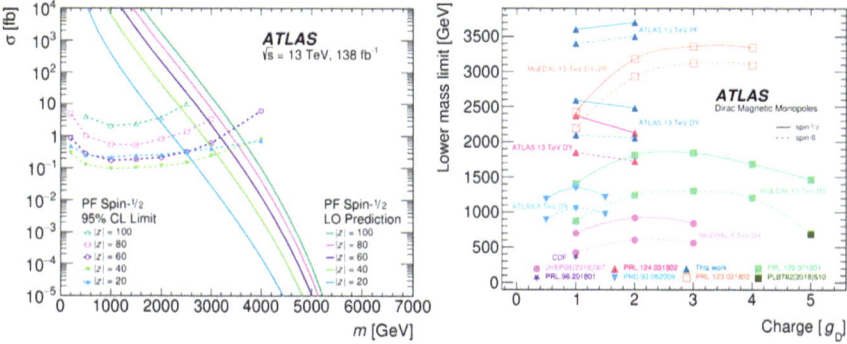

Fig. 5 The observed 95% CL upper limits on the cross section for all masses and charges of photon-fusion (bottom) pair-produced HECOs for spin-1/2 (left); Comparison of the lower mass limits obtained by LHC searches in Run 1 and Run 2 pp collisions for Drell-Yan and photon-fusion pair-produced magnetic monopoles (right). Reprinted under CC-BY-4.0 license from [11]. © 2024, CERN for the ATLAS Collaboration

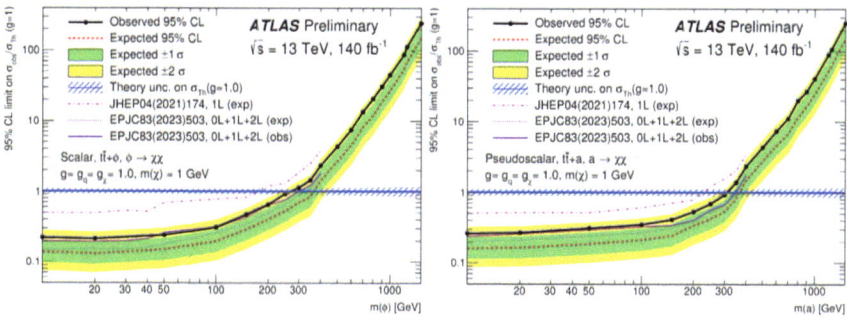

Fig. 6 Upper limit at 95% CL on the ratio of the $t\bar{t}$+DM production cross-section to the theoretical cross-section under the hypothesis of (left) a scalar or (right) a pseudoscalar mediator. Limits are shown as a function of $m(\Phi/a)$ assuming $m(\chi) = 1$ GeV. Reprinted under CC-BY-4.0 license from [12]. © 2024, CERN for the benefit of ATLAS Collaboration

simplified benchmark model for the $t\bar{t}$+DM production, it assumes the existence of a (pseudo)scalar mediator Φ, and Φ can be produced in association with two top quarks and decay into a pair of SM-singlet DM particles (χ).

The search for top quark pairs with the ATLAS Run 2 datasets is performed in final states with one lepton, jets and missing transverse momentum [12]. The processes $t\bar{t}$, single-top (including tW) and W+jets are the dominant sources of background in this analysis. As shown in Fig. 6, the exclusion limits at 95% confidence level for the $t\bar{t}$+DM production are set. No significant excess above the Standard Model background is observed.

7 Conclusions

There are still many open questions to be understood in our universe and that the answers lie beyond the standard model. Therefore, it's crucial to search for new phenomena at the LHC. The latest BSM physics searches with the ATLAS detector have been enhanced by the large dataset delivered by the LHC during the Run 2 and the dedicated analyses performed targeting specific BSM signatures.

Despite the extensive analyses performed and the various tested models, no evidence for new physics has been found so far. Consequently, improved mass limits have been set for theoretical particles within the respective benchmark models and the exclusion range of the models phase-space has been extended. The prospects for BSM physics searches remain promising for Run 3 and the High Luminosity LHC, as they will benefit from larger datasets, as well as upgraded detector and trigger capabilities.

References

1. ATLAS Collaboration, The ATLAS experiment at the CERN Large Hadron Collider. J. Instrum. **3**, S08003 (2008)
2. ATLAS Collaboration, Search for pair production of higgsinos in events with two Higgs bosons and missing transverse momentum in $\sqrt{s} = 13$ TeV pp collisions at the ATLAS experiment. Phys. Rev. D **109**, 112011 (2024)
3. ATLAS Collaboration, Search for direct production of electroweakinos in final states with one lepton, jets and missing transverse momentum in pp collisions at $\sqrt{s} = 13$ TeV with the ATLAS detector. J. High Energy Phys. **12**, 167 (2023)
4. ATLAS Collaboration, Search for quantum black hole production in lepton+jet final states using proton-proton collisions at $\sqrt{s} = 13$ TeV with the ATLAS detector. Phys. Rev. D **109**, 032010 (2024)
5. ATLAS Collaboration, Pursuit of paired dijet resonances in the Run 2 dataset with ATLAS. Phys. Rev. D **108**, 112005 (2023)
6. ATLAS Collaboration, A search for R-parity-violating supersymmetry in final states containing many jets in $\sqrt{s} = 13$ TeV pp collisions with the ATLAS detector. J. High Energy Phys. **2405**, 003 (2024)
7. ATLAS Collaboration, Search for vector-boson resonances decaying into a top quark and a bottom quark using pp collisions at $\sqrt{s} = 13$ TeV with the ATLAS detector. J. High Energy Phys. **12**, 073 (2023)
8. ATLAS Collaboration, Search for pair-production of vector-like quarks in lepton+jets final states containing at least one b-jet using the Run 2 data from the ATLAS experiment. Phys. Lett. B **854**, 138743 (2024)
9. ATLAS Collaboration, Search for single vector-like B quark production and decay via $B \rightarrow bH(b\bar{b})$ in pp collisions at $\sqrt{s} = 13$ TeV with the ATLAS detector. J. High Energy Phys. **11**, 168 (2023)
10. A. Polyakov, Particle spectrum in the quantum field theory. JETP Lett. **20**, 194 (1974)
11. ATLAS Collaboration, Search for magnetic monopoles and stable particles with high electric charges in $\sqrt{s} = 13$ TeV pp collisions with the ATLAS detector. J. High Energy Phys. **11**, 112 (2023)

12. ATLAS Collaboration, Search for new phenomena with top-quark pairs in final states with one lepton, jets and missing transverse momentum using 140 fb^{-1} of data at $\sqrt{s} = 13$ TeV with the ATLAS detector. J. High Energy Phys. **2403**, 139 (2024)
13. L. Evans, P. Bryant, LHC machine. J. Instrum. **3**, S08001 (2008)

Open Access This chapter is licensed under the terms of the Creative Commons Attribution 4.0 International License (http://creativecommons.org/licenses/by/4.0/), which permits use, sharing, adaptation, distribution and reproduction in any medium or format, as long as you give appropriate credit to the original author(s) and the source, provide a link to the Creative Commons license and indicate if changes were made.

The images or other third party material in this chapter are included in the chapter's Creative Commons license, unless indicated otherwise in a credit line to the material. If material is not included in the chapter's Creative Commons license and your intended use is not permitted by statutory regulation or exceeds the permitted use, you will need to obtain permission directly from the copyright holder.

Recent Results from Belle and Belle II Experiments

Paolo Branchini

1 Belle II and SuperKEKB

Belle II represents a substantial upgrade from the original Belle experiment, offering enhanced performance and increased data acquisition capabilities, as discussed in reference [1]. Situated at the KEK laboratory in Tsukuba, Japan, Belle II operates in conjunction with the SuperKEKB e^+e^- collider. Its research objectives encompass a wide range of topics, including explorations of the dark sector, the study of rare and prohibited B meson decays, investigations into lepton flavor and CP asymmetries, as well as examinations of charm and tau physics, as detailed in [2]. Notably, Belle II has accumulated 428 fb^{-1} of data since March 2019. Currently, the facility is undergoing an extended period of inactivity known as long shutdown 1, which spans from July 2022 to September 2023. This downtime is dedicated to the installation of a new two-layer pixel vertex detector. SuperKEKB ranks as the world's highest instantaneous luminosity collider, achieving a peak luminosity of 4.7×10^{34} cm^{-2} s^{-1}. To attain the ambitious target of 6×10^{35} cm^{-2} s^{-1}, efforts within the accelerator development domain are concentrated on increasing the current while simultaneously reducing injection-related backgrounds. Additional priorities involve minimizing catastrophic beam loss events and managing emittance blowup and beam instability. An international task force has been established to provide insights, including proposals for potential upgrades to the final focus or other hardware improvements during long shutdown 2, slated for 2028.

P. Branchini (✉)
RomaTre INFN Division, Rome, Italy
e-mail: paolo.branchini@roma3.infn.it

2 Dark Sector

We compute upper limits for the coupling constant g' as a function of the Z' mass within the $L\mu - L\tau$ framework, assuming $B(Z' \rightarrow \chi \chi) = 1$. These findings, illustrated in the right panel of Fig. 1, rule out the parameter space favored by the $(g-2)_\mu$ anomaly for Z' [3] masses ranging from 0.8 to 5.

In a similar vein, we search for a distinctive feature in the mass distribution of the recoil observed when two muons interact with an additional τ pair reconstructed from τ decays into a single charged particle, $\tau \rightarrow h, \ell\nu(\nu)$. To accomplish this, we leverage the unique characteristics of the signal, including final state radiation and di-tau resonant processes. This is done through the application of multi-layer perceptron neural networks to effectively suppress QED background, a process carried out independently in eight distinct mass regions.

To mitigate the impact of known inaccuracies in the simulation, we directly fit the expected background using the recoil mass distribution from the data. Upon analyzing 63 fb^{-1} of data, no statistically significant excess is observed when fitting the recoil mass range within the range of $3.6 < M_{recoil(\mu\mu)} < 10$ GeV/c^2, employing half mass resolution steps. As a result, we establish upper limits at a 90% confidence level for the quantity $\sigma(e^+e^- \rightarrow \mu^+\mu^-\tau^+\tau^-) \times B(X \rightarrow \tau^+\tau^-)$, which can be interpreted across various model classes. This search establishes the world's most stringent limits on the scalar coupling constant ξ within leptophilic scalar models [4] for $m_s > 6.5$ GeV/c^2, as depicted in the left panel of Fig. 1 in the right panel limits on the coupling constant of the axion model vs the mass of the axion like particles are shown.

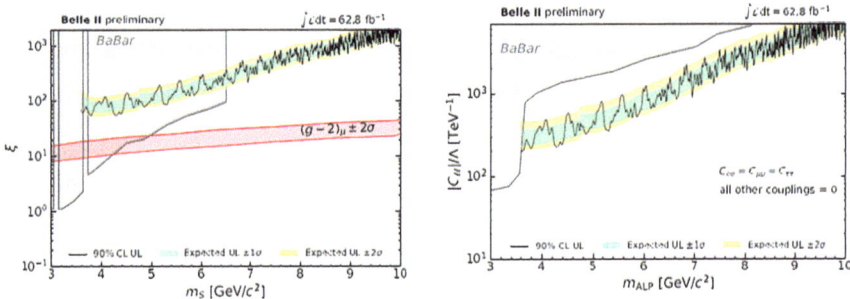

Fig. 1 The 90% CL upper limits on the leptophilic coupling constant ξ as a function of the searched MS are shown on the left panel. On the right one limits of the coupling constant of axion like particles are shown vs the Mass, assuming universality on lepton couplings

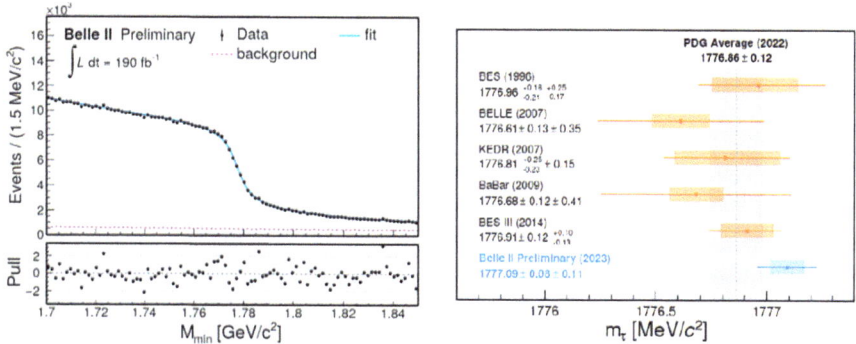

Fig. 2 On the left, the spectrum of the reconstructed pseudomass in data (black dots) and the superimposed fit (solid blue line) are shown. The bottom inset plot displays differences between data and fit result divided by the statistical uncertainties. On the right, the summary of the most precise measurements of the tau mass to date, compared to the world average (gray band) and this work result (blue text)

3 Tau Mass Measurements

Lepton properties represent fundamental parameters within the Standard Model (SM) that necessitate the highest degree of precision in measurement. Belle II is well-suited for the accurate determination of several properties of τ leptons. Utilizing the pseudo-mass $M_{\min}$ technique on reconstructed $e^+e^- \to \tau^+\tau^-$ events from a dataset of 190 fb^{-1}, we present the most precise measurement in the world of the τ lepton mass, denoted as $M\tau$. The extracted value is obtained through a fit to the endpoint of the distribution $M_{\min} = \sqrt{M^2_{3\pi} + 2(\sqrt{s}/2 - E^*_{3\pi})(E^*_{3\pi} - P^*_{3\pi})}$. This quantity is computed from events in which one of the τ leptons decays into three charged pions while the other τ lepton decays into a single charged particle. The distributions of the pseudo-mass in both simulation and data are displayed in the left panel of Fig. 2. Achieving the highest precision in this measurement necessitates stringent control of systematic uncertainties, primarily stemming from the calibration of beam energies and the scaling of charged-particle momenta. This control reduces the overall systematic uncertainty to 0.11 MeV/c^2. As a result, we achieve the most accurate measurement to date of the τ lepton mass, yielding a value of 1777.09 $\pm$ 0.08stat $\pm$ 0.11sys [1].

4 τ Michel parameter ξ'

We performed an event selection process for $\tau^- \to \mu^- (\to e^- \nu_e \nu_\mu) \nu_\mu \nu_\tau$ decays based on the characteristics of $\tau^+\tau^-$-pair event topology, kink patterns, and $\mu^- \to e^- \nu_e \nu_\mu$ decay features. The primary focus was on selecting events where the $\tau^- \to$

$\mu^- \nu_\mu \nu_\tau$ decay occurred on the signal side, without a specific concern for the decay mode of the tagging τ lepton. To achieve this, they used the event topology criteria of 1–1 or 1–3, which represented one charged track originating from the beam interaction point (IP) in the signal hemisphere and one or three charged tracks from the IP in the tag hemisphere. In the signal hemisphere, an additional requirement involved one track displaced from the IP, serving as an electron candidate originating from the $\mu^- \to e^- \nu_e \nu_\mu$ decay. To suppress non-$\tau^+\tau^-$ background events such as hadron and muon production, two-photon processes, and Bhabha scattering, they relied on the characteristic missing energy and momentum signature associated with $\tau^+\tau^-$ pair events. Bhabha scattering was further reduced through an electron veto in the tagging hemisphere using standard Belle particle identification techniques. Additionally, background events in the signal hemisphere arising from decay modes other than $\tau^- \to \mu^- \nu_\mu \nu_\tau$ were minimized by rejecting events with significant energy deposits in the electromagnetic calorimeter (ECL) related to photons in the signal hemisphere. The innovative aspect of this analysis involved the use of kinks, which had not been employed in previous Belle studies. Selection criteria were based on two reconstructed tracks in the CDC: one originating near the IP (muon candidate) and ending inside the CDC volume, and the other track (electron candidate) was required to be inconsistent with production at the IP. Both tracks needed to be in close proximity, forming a kink, with the point of their closest approach used as the muon decay vertex. The kink trajectories were typically shorter than regular trajectories, and this was used to limit the number of CDC hits assigned to the tracks. Further background suppression was achieved by vetoing signals in Belle subsystems that were absent for kink tracks, such as TOF, ECL, and KLM response for the muon track candidate, and SVD and KLM response for the electron candidate. After implementing these two selection steps, nonkink background events were reduced to a negligible level, as confirmed using Monte Carlo (MC) simulation. The remaining events consisted of the signal $\mu^- \to e^- \nu_e \nu_\mu$ decay and real-kink background processes that mimicked the signal. These background processes included light meson decays and electron, muon, and hadron scattering. Backgrounds from K− and π − decays were mainly two-body and had characteristic monochromatic daughter particle momentum when the correct pair of mass hypotheses was assigned to the tracks. To further suppress the remaining background with high efficiency, a boosted decision tree classifier (BDT) was employed. The BDT was configured to distinguish signal from background based on 12 selected features, including momentum variables for daughter particles in the mother rest frame, particle identification variables for muon and electron candidates, and parameters of the decay vertex. With the BDT selection, a high signal efficiency (around 80%) was achieved while suppressing background by a factor of 50. The MC simulation confirmed that the BDT selection did not introduce any bias into the signal kinematic distributions. To measure the Michel parameter ξ^0, an unbinned maximum-likelihood fit was performed on the distribution of (y, cos θ_e), where y represented the electron energy in the muon rest frame divided by half the muon mass, and θe was an angle that provided information about the muon spin. The fit was based on signal and background probability density functions (PDFs), and the Michel parameter was determined to be $\xi^0 = 0.22 \pm 0.94$(stat) ± 0.42(sys), consistent with the Standard Model expectation of $\xi^0 = 1$ within the uncertainties. In summary, this study presented the first direct measurement of the Michel parameter ξ^0 in $\tau^- \to \mu^- \nu_\mu \nu_\tau$ decays using a comprehensive data sample. The measured value, $\xi^0 = 0.22 \pm 0.94 \pm 0.42$, aligned with the SM prediction and demonstrated significantly improved precision compared to previous measurements.

5 Search for $B^+ \to K^+ + \nu\nu$ Decays

The search for the $B^+ \to K^+ + \nu\nu$ decay presents a unique opportunity within the Belle II experiment. This decay mode has never been observed previously, and the Standard Model amplitude [4] can potentially receive significant contributions from Beyond the Standard Model (BSM) amplitudes. The measurement is carried out using a dataset with an integrated luminosity of 63 fb^{-1}. Reconstruction is accomplished through an inclusive tagging approach, where the B_{sig} particle is reconstructed by identifying the highest p_T track that is consistent with a K^+ meson, while the remainder of the event is assigned to the B_{tag}.

This procedure is validated using decays such as $B^+ \to J/\psi\ (\to \mu\mu)\ K^+$ as benchmarks. To enhance the purity of the signal and reduce background, a two-stage Boosted Decision Tree (BDT) approach is applied, leveraging event shape, kinematic properties, and vertex characteristics. No signal is observed in the analysis, and the resulting upper limit is determined to be 4.1×10^{-5} at a 90% confidence level. The result is also expressed in terms of signal strength or Br($B^+ \to K^+ + \nu\nu$) = $(1.9 \pm 1.3$ (stat) $+0.08 - 0.07$ (syst)) $\times 10^{-5}$, a value consistent with both the Standard Model prediction and previous findings [5]. Projections based on larger datasets suggest that a 5σ observation can be achieved with an integrated luminosity of 5 ab^{-1}, accompanied by an expected 50% improvement in efficiency due to the use of exclusive tagging methods in conjunction with the inclusive one. Furthermore, additional decay channels (such as K* K^0_S) will be explored in future investigations.

References

1. T. Abe (Belle II Collaboration), Belle II Technical Design Report.https://doi.org/10.48550/arXiv.1011.0352 (2010), arXiv:1011.0352
2. Belle II Collaboration, Snowmass White Paper: Belle II physics reach and plans for the next decade and beyond, arXiv:2207.06307
3. B. Batell, N. Lange, D. McKeen, M. Pospelov, A. Ritz, Phys. Rev. D. **95**, 075003 (2017). https://doi.org/10.1103/PhysRevD.95.075003
4. A.J. Buras, J. Girrbach-Noe, C. Niehoff, D.M. Straub, B $\to$ K^+ $\nu\nu$ decays in the standard model and beyond. J. High Energy Phys. **02**, 184 (2016). https://doi.org/10.1007/JHEP02(2016)184

5. W.G. Parrott, C. Bouchard, C.T.H. Davies (HPQCD Collaboration), Standard Model predictions for $B \to K\ell^+\ell^-$, $B \to K\ell_1^-\ell_2^-$, and $B \to K\nu\bar{\nu}$ using form factors from $N_f = 2+1+1$ lattice QCD, Phys. Rev. D **107**, 014511 (2023). https://doi.org/10.1103/PhysRevD.107.014511

Open Access This chapter is licensed under the terms of the Creative Commons Attribution 4.0 International License (http://creativecommons.org/licenses/by/4.0/), which permits use, sharing, adaptation, distribution and reproduction in any medium or format, as long as you give appropriate credit to the original author(s) and the source, provide a link to the Creative Commons license and indicate if changes were made.

The images or other third party material in this chapter are included in the chapter's Creative Commons license, unless indicated otherwise in a credit line to the material. If material is not included in the chapter's Creative Commons license and your intended use is not permitted by statutory regulation or exceeds the permitted use, you will need to obtain permission directly from the copyright holder.

Measurements of Heavy-flavour Production in pp Collisions with the ALICE Detector

Tebogo Joyce Shaba

1 Introduction

Heavy quarks (charm and beauty) are abundantly produced at the LHC [13] in the early stages of hadronic collisions via hard parton scattering processes and pertubative QCD (pQCD) calculations can be used to estimate their production cross sections. Heavy-quark production is experimentally accessible through the measurement of heavy-flavour hadrons and their decay products. The measurement of heavy flavours in pp collisions can be thus used to test pQCD calculations. In addition, the ratios of heavy-flavour production yields, measured for different beam energies and different rapidity intervals, are key observables that are sensitive to the heavy-quark fragmentation functions. Such ingredients of the heavy-flavour production cross section calculations have recently been observed to be significantly collision-system dependent.

ALICE Collaboration.
Tebogo Joyce Shaba for the ALICE collaboration.

T. J. Shaba (✉)
North-West University, North-West, South-Africa

iThemba LABS, Somerset West, Western Cape, South-Africa
e-mail: tebogo.joyce.shaba@cern.ch

2 Results

2.1 Production Cross Sections of Leptons from HF-hadron Decays

The inclusive production cross section measurement of heavy-flavour hadron decay muons performed at $\sqrt{s}$ = 5.02 TeV [2] is compared to FONLL [9] pQCD calculations in Fig. 1a and the inclusive production cross section measurement of heavy-flavour hadron decay electrons performed at $\sqrt{s}$ = 13 TeV [4] is compared to FONLL [9] and GM-VFNS [8] pQCD calculations in Fig. 1b. At both energies the data lie on the upper edge of the FONLL predictions. The GM-VFNS calculation underestimates the cross section at low p_T but describes the data within the uncertainties for $p_T > 5$ GeV/c. The production cross section of heavy-flavour hadron decay leptons at low p_T mainly probes charm production while at high p_T it mainly probes the beauty production.

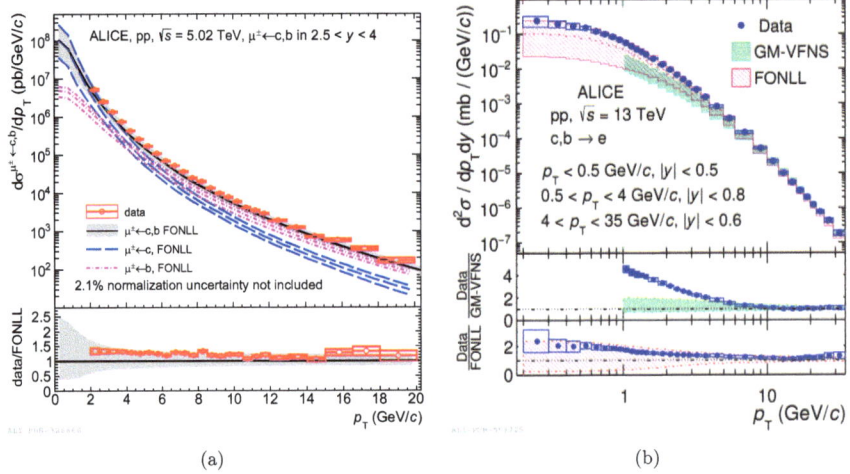

Fig. 1 Two-panel figure: (**a**) [2] and (**b**) [4] showing X–Y charts related to particle physics data from the ALICE experiment. (**a**) displays the differential cross section $d\sigma^{\mu^+\mu^-\to c,b}/dp_T$ as a function of transverse momentum pT (in GeV/c). Data points are shown as red squares, and theoretical predictions from FONLL are represented with various dashed lines. A note indicates that a 2.1% normalization uncertainty is not included. A lower subplot shows the ratio of data to the FONLL predictions. (**b**) presents the differential cross section $d^2\sigma/dp_T dy$ for charm and bottom quarks decaying to electrons, as a function of pT (in GeV/c). The data are compared with GM-VFNS and FONLL model predictions, shown with different colored bands. The chart includes specific transverse momentum and rapidity ranges, $|y|$. A lower subplot displays the ratio of data to model predictions. The measurements are compared with pQCD predictions

2.2 Prompt Charm-hadron Production Yield Ratios

Figure 2 shows the ratios of production yields of prompt D mesons, namely D^+/D^0 and D_s^+/D^0 as a function of p_T in pp collisions at $\sqrt{s}$ = 5.02, 7 and 13 TeV [6]. The measurements and their ratios are used to determine the charm-quark fragmentation fractions [5, 6]. The production yield ratios do not show any significant dependence on p_T or centre-of-mass energy. These results suggest a common fragmentation function of charm quarks into D-mesons at different LHC energies. Figure 3a shows the ratio between the production yields at midrapidity of prompt Λ_c^+ baryons and D^0 mesons in pp collisions at $\sqrt{s}$ = 5.02, 7 and 13 TeV [6] compared with predictions from PYTHIA Monash tune [17], PYTHIA CR-BLC Mode 0, 2 and 3 [11], SHM+RQM [15], Catania [16] and QCM [18]. The ratio of the production yield times the branching fraction of the heavier charm baryon Ω_c^0 over the production yield of D^0 as a function of p_T in pp collisions at $\sqrt{s}$ = 13 TeV are shown in Fig. 3b [7]. The Λ_c^+/D^0 ratio is underestimated by PYTHIA8 Monash by a factor of 2–5 depending on the p_T while the Ω_c^0/D^0 ratio is underestimated by several orders of magnitude. PYTHIA 8 with CR-BLC models and the SHM-RQM reproduce correctly the Λ_c^+/D^0 ratio, but underestimate the Ω_c^0/D^0. The catania model describes both measurements well. QCM underestimates Ω_c^0/D^0 ratios, while correctly reproducing the Λ_c^+ ratios.

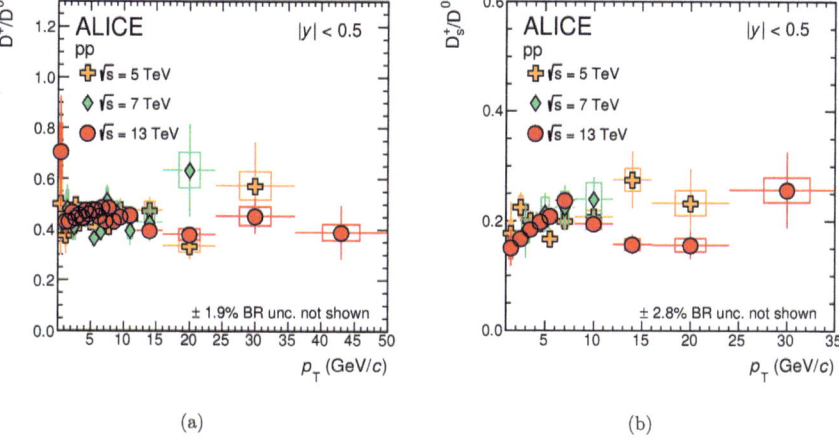

Fig. 2 Two-panel X–Y chart showing particle physics data from the ALICE experiment for production yields as a function of p_T of prompt D^+/D^0 (**a**) and D_s^+/D^0 (**b**) mesons in pp collisions, with data points for $\sqrt{s}$ = 5, 7, and 13 TeV, represented by orange crosses, green diamonds, and red circles, respectively. The vertical axis ranges from 0 to 1.2, and the horizontal axis spans from 0 to 50 GeV/c (**a**) while the vertical axis ranges from 0 to 0.6, and the horizontal axis spans from 0 to 3.5 GeV/c (**b**) [5]. Both panels note a branching ratio uncertainty that is not shown. The rapidity range is $|y| < 0.5$

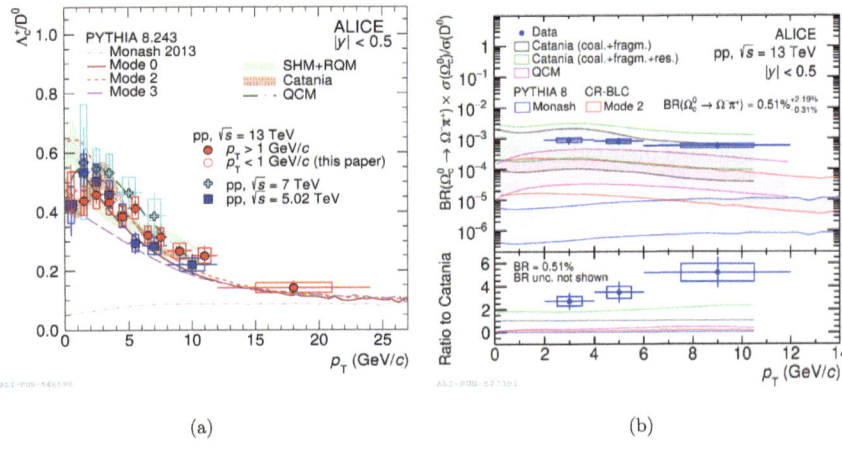

Fig. 3 Panel figure showing two X–Y charts related to particle physics data from ALICE experiments for the p_T-differential production yield ratios at midrapidity of prompt Λ_c^+ baryons and D^0 mesons. (**a**) shows the ratio of Λ_c to D^0 particles as a function of transverse momentum p_T (in GeV/c). Data points are shown for different collision energies: 13, 7, and 5.02 TeV. Theoretical predictions are included from models such as PYTHIA 8.243 (Monash 2013) [17] and others. Shaded bands represent additional model calculations including SHM+RQM [15], Catania [16], and QCM [18]. (**b**) displays the product of the branching ratio of $\Omega_c \to \Xi_\pi$ and the ratio Ω_c/D^0, plotted as a function of p_T (in GeV/c). The chart includes both data and predictions from the Catania model, PYTHIA 8, and CR-BLC. A box plot shows the ratio to the Catania prediction, with a noted branching ratio of 0.51 %. Both charts are labeled with ALICE and include legends and annotations for clarity

2.3 Multiplicity Dependence of Charm-hadron Ratios

The measurement of heavy-quark production as a function of charged-particle multiplicity allows us to investigate the interplay between the hard and soft particle-production processes, the role of multiparton interactions, and the effect of color-reconnection mechanisms. Figure 4 shows the D_s^+/D^0 (top) and Λ_c^+/D^0 (bottom) ratios measured in pp collisions at $\sqrt{s} = 13$ TeV for the low (left) and high (right) multiplicity classes at midrapidity [3]. The measured D_s^+/D^0 ratios do not show a p_T or multiplicity dependence and are compatible with PYTHIA8 with Monash and CR-BLC tunes. The CE-SH model [10] is compatible with D_s^+/D^0 for low multiplicity events, while it overestimates the data at high multiplicity. On the other hand, the Λ_c^+/D^0 ratio shows both a p_T and a multiplicity dependence, with an increase in the ratio at higher charged-particle multiplicity. PYTHIA8 + Monash tune [17] does not reproduce the Λ_c^+/D^0 ratio and, furthermore, it does not show any multiplicity dependence. When using CR-BLC modes [12], the Λ_c^+/D^0 ratios are, however, well described. The ratios of Λ_c^+/D^0 show an enhancement when compared to e^+e^- collisions, while D_s^+/D^0 is consistent with that measured in e^+e^- collisions [14] (Fig. 5).

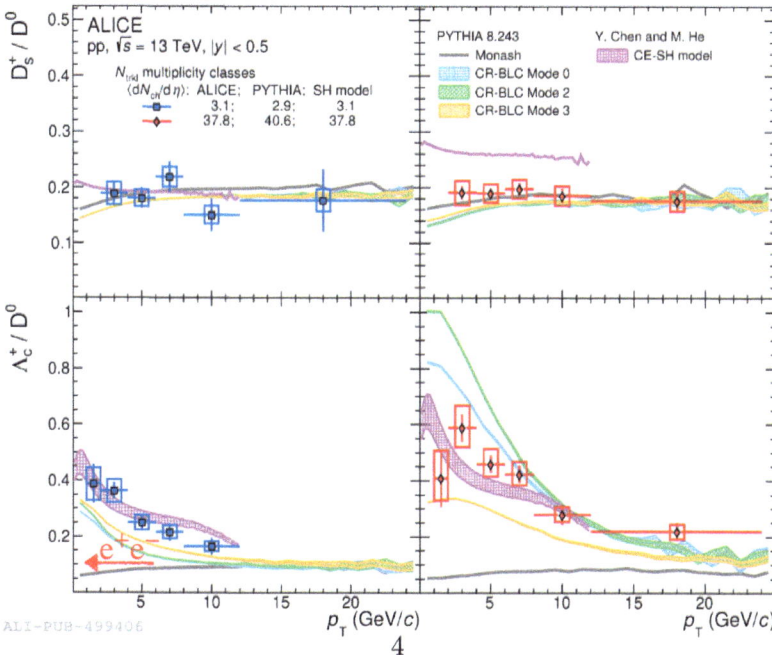

Fig. 4 The panel figure shows the D_s^+/D^0 (top) and Λ_c^+/D^0 (bottom) ratios measured in pp collisions at $\sqrt{s} = 13$ TeV. All ratios are plotted as a function of transverse momentum p_T (in GeV/c). The left panels present data from the ALICE experiment, including different charged-particle multiplicity classes. The right panels compare the experimental results to model predictions from PYTHIA [17] and the CE-SH [10] framework. Various lines and markers represent different models and experimental data, with legends indicating CR-BLC [12] configurations and other model details. The figure includes both statistical and systematic uncertainties. [3]

3 Conclusion

The measurements of the production cross sections of muons from HF-hadron decays at forward-rapidity are in agreement with FONLL predictions. The p_T-differential ratios of the measured charm-hadron cross sections suggest a common fragmentation functions of charm into hadrons at different LHC energies. The ratios of Λ_c^+/D^0 show an enhancement when compared to e^+e^- collisions and a charged particle-multiplicity dependence, not observed for the D_s^+/D^0 case. The comparison of these measurements with model predictions suggests that a modified hadronisation with respect to in-vacuum fragmentation is needed to correctly describe the results. These self-normalised yields allow a direct comparison of multiplicity-dependent production of different particle species. A strong p_T dependence is observed for electrons, with high-p_T electrons showing a faster-than-liner increase as a function of the self-normalised multiplicity. With the significantly

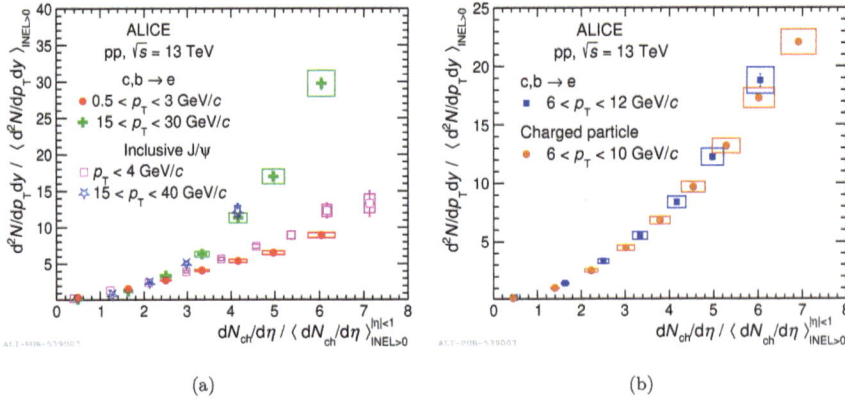

Fig. 5 Two-panel showing the comparison of the self-normalised yields of electrons from heavy-flavour hadron decays from the ALICE experiment at $\sqrt{s} = 13$ TeV. (**a**) displays the normalized differential cross section for charm and beauty quarks decaying to electrons, as well as inclusive J/ψ production. Various transverse momentum ranges are indicated by distinct symbols and colors. (**b**) presents similar measurements for charm and beauty quarks, and for charged particles, again using different symbols and colors to represent different momentum intervals. In both panels, the normalized yield is plotted as a function of the charged-particle multiplicity density [4]

higher integrated luminosity expected to be collected during LHC Run 3 compared to Runs 1 and 2, ALICE will enable more precise measurements of heavy-flavor production, further advancing our understanding of these phenomena.

References

1. ALICE Collaboration, Measurement of open heavy-flavour production as a function of charged-particle multiplicity with ALICE at the LHC. Int. J. Mod. Phys. A **20**, 1430044 (2014). https://doi.org/10.1142/S0217751X14300440
2. ALICE Collaboration, First observation of an attractive interaction between a proton and a cascade baryon. J. High Energy Phys. **2019**, 008 (2019). https://doi.org/10.1103/PhysRevLett.123.112002
3. ALICE Collaboration, Observation of a multiplicity dependence in the p_T-differential charm baryon-to-meson ratios in proton–proton collisions at $\sqrt{s} = 13$ TeV. Phys. Lett. B **829**, 137065 (2022). https://doi.org/10.1103/PhysRevLett.131.061901
4. ALICE Collaboration, Prompt and non-prompt J/ψ production at midrapidity in Pb–Pb collisions at $\sqrt{s_{NN}} = 5.02$ TeV. J. High Energy Phys. **2023**, 006 (2023). https://doi.org/10.1007/JHEP02(2024)066
5. ALICE Collaboration, Measurements of long-range two-particle correlation over a wide pseudorapidity range in p–Pb collisions at $\sqrt{s_{NN}} = 5.02$ TeV. J. High Energy Phys. **2023**, 086 (2023). https://doi.org/10.1007/JHEP12(2023)086
6. ALICE Collaboration, Evidence for modification of b quark hadronization in high-multiplicity pp collisions at $\sqrt{s} = 13$ TeV. J. High Energy Phys. **2023**, 086 (2023). https://doi.org/10.1007/JHEP12(2023)086

7. ALICE Collaboration, First measurement of Ω_c^0 production in pp collisions at $\sqrt{s} = 13$ TeV. Phys. Lett. B **846**, 137625 (2023). https://doi.org/10.1016/j.physletb.2023.137625
8. P. Bolzoni, G. Kramer, Inclusive charmed-meson production from bottom hadron decays at the LHC. J. Phys. G Nucl. Part. Phys. **41**(7), 075006 (2014). https://doi.org/10.1088/0954-3899/41/7/075006
9. M. Cacciari, S. Frixione, N. Houdeau, et al., Theoretical predictions for charm and bottom production at the LHC. J. High Energy Phys. **2012**(10), 1–24. https://doi.org/10.1007/JHEP10(2012)137
10. Y. Chen, M. He, Charged-particle multiplicity dependence of charm-baryon-to-meson ratio in high-energy proton–proton collisions. Phys. Lett. B **815**, 136144 (2021). https://doi.org/10.1016/j.physletb.2021.136144
11. R. Christiansen, P.Z. Skands, String formation beyond leading colour. J. High Energy Phys. **2015**, 003 (2015). https://doi.org/10.1007/JHEP08(2015)003
12. R. Christiansen, P.Z. Skands, VBF vs. GGF Higgs with full-event deep learning: Towards a decay-agnostic tagger. J. High Energy Phys. **2015**(8), 041 (2015). https://doi.org/10.1007/JHEP08(2015)041
13. L. Evans, P. Bryant, LHC machine. J. Instrum. **3**, S08001 (2008). https://doi.org/10.1088/1748-0221/3/08/S08001
14. L. Gladilin, Fragmentation fractions of c and b quarks into charmed hadrons at LEP. Eur. Phys. J. C **75**(1), 19 (2015). https://doi.org/10.1140/epjc/s10052-014-3024-y
15. M. He, R.J. Fries, R. Rapp, Ω_{ccc} production in high-energy nuclear collisions. Phys. Lett. B **1**(24), 2012 (2012). https://doi.org/10.1016/j.physletb.2012.03.031
16. V. Minissale, S. Plumari, V. Greco, Charm hadrons in pp collisions at LHC energy within a coalescence plus fragmentation approach. Phys. Lett. B **821**, 136622 (2021). https://doi.org/10.1016/j.physletb.2021.136622
17. P.Z. Skands, S. Carrazza, J. Rojo, Tuning PYTHIA 8.1: the Monash 2013 tune. Eur. Phys. J. C **74**(8), 3024 (2014). https://doi.org/10.1140/epjc/s10052-014-3024-y
18. J. Song, H. Li, F. Shao, New feature of low p_T charm quark hadronization in pp collisions at $\sqrt{s} = 7$ TeV. Eur. Phys. J. C **78**, 1–8 (2018). https://doi.org/10.1140/epjc/s10052-018-6043-6

Open Access This chapter is licensed under the terms of the Creative Commons Attribution 4.0 International License (http://creativecommons.org/licenses/by/4.0/), which permits use, sharing, adaptation, distribution and reproduction in any medium or format, as long as you give appropriate credit to the original author(s) and the source, provide a link to the Creative Commons license and indicate if changes were made.

The images or other third party material in this chapter are included in the chapter's Creative Commons license, unless indicated otherwise in a credit line to the material. If material is not included in the chapter's Creative Commons license and your intended use is not permitted by statutory regulation or exceeds the permitted use, you will need to obtain permission directly from the copyright holder.

Part II
Posters

The W Boson Mass Anomaly within the Inverted Scenario of the Two-Higgs Doublet Model

Hamza Abouabid, Abdesslam Arhrib, Rachid Benbrik ⓘ, Mohamed Krab, and Mohamed Ouchemhou

1 Introduction

Recent findings from the CDF collaboration, utilizing data from the Tevatron, revealed a significant W boson mass of $M_W^{CDF} = 80.4435 \pm 0.0094\,\text{GeV}$ [1], contrasting the SM prediction of $M_W^{SM} = 80.357 \pm 0.006\,\text{GeV}$ [2, 3]. This deviation, with a 7σ significance, suggests new physics beyond the Standard Model, particularly in scenarios with extended Higgs sectors.

In light of this outcome, we explore the implications of the CDF measurement within the 2HDM, which includes a spectrum of two CP-even Higgs bosons (h and H), one CP-odd A, and a pair charged Higgs bosons $H^\pm$ [4–6]. We identify the observed SM Higgs with the heavier CP-even Higgs H. Our focus is on how Two-Higgs Doublet Model (2HDM) may reconcile the M_W^{CDF} discrepancy and the resulting implications for Higgs boson decays in 2HDM type-I.

The structure of our contribution is concise and focused. In Sect. 2, we discuss the 2HDM setup, including the computation of M_W^{2HDM} and $\sin^2\theta_{\text{eff}}^{2HDM}$. Section 3 covers the parameter space analysis and our key findings on the W boson mass within the 2HDM framework. We conclude in Sect. 4.

H. Abouabid (✉) · A. Arhrib
Abdelmalek Essaadi University, Faculty of Sciences and Techniques, Tangier, Morocco
e-mail: hamza.abouabid@etu.uae.ac.ma

R. Benbrik · M. Ouchemhou
Laboratory of Fundamental and Applied Physics, Faculté Polydisciplinaire de Safi, Sidi Bouzid, Safi, Morocco

M. Krab
Department of Physics, National Taiwan University, Taipei, Taiwan

© The Author(s) 2026
Y. Tayalati, M. Gouighri (eds.), *The First African Conference on High Energy Physics*, Springer Proceedings in Physics 425,
https://doi.org/10.1007/978-3-031-88933-2_31

2 M_W Within the 2HDM

The 2HDM extends the SM by introducing a second Higgs doublet. This addition enriches the scalar sector and enhances collider phenomenology. Its Higgs potential is given by:

$$V_{2HDM} = m_1^2(\phi_1^\dagger \phi_1) + m_2^2(\phi_2^\dagger \phi_2) - [m_{12}^2(\phi_1^\dagger \phi_2) + h.c.]$$
$$+ \frac{\lambda_1}{2}(\phi_1^\dagger \phi_1)^2 + \frac{\lambda_2}{2}(\phi_2^\dagger \phi_2)^2$$
$$+ \lambda_3(\phi_1^\dagger \phi_1)(\phi_2^\dagger \phi_2) + \lambda_4(\phi_1^\dagger \phi_2)(\phi_2^\dagger \phi_1) + \frac{1}{2}[\lambda_5(\phi_1^\dagger \phi_2)^2 + h.c.], \quad (1)$$

where λ_{1-5}, m_1^2 and m_2^2 are real parameters. After Electro-Weak (EW) symmetry breaking, the 2HDM consists of seven free parameters that we have chosen as: M_h, M_H, M_A, $M_{H^\pm}$, $\sin(\beta - \alpha)$, $\tan\beta$ and m_{12}^2.

In order to suppress Flavor Changing Neutral Currents (FCNCs) at the tree level, a discrete Z_2 symmetry is imposed, leading to four types of the model based on Yukawa interactions. In 2HDM type-I, both up- and down-type quarks and leptons couples to ϕ_2, while in type-II up-type quarks couples to ϕ_2 and down-type quarks and leptons interacts with ϕ_1. In type-X (or Lepton-specific) assigns quarks to ϕ_2 and leptons to ϕ_1. Conversely, type-Y (or Flipped) couples up-type quarks and leptons to ϕ_2, and down-type quarks to ϕ_1.

The contributions of the 2HDM to the EW Precision Observables (EWPOs) are encapsulated by the oblique parameters S, T, and U [7]. The impact of these contributions on the W boson mass and the effective weak mixing angle is expressed as:

$$\Delta M_W^2 = \frac{\alpha_0 c_W^2 M_Z^2}{c_W^2 - s_W^2}\left[-\frac{1}{2}S + c_W^2 T + \frac{c_W^2 - s_W^2}{4s_W^2}U\right], \quad (2)$$

$$\Delta \sin^2\theta_{eff} = \frac{\alpha_0}{c_W^2 - s_W^2}\left[\frac{1}{4}S - s_W^2 c_W^2 T\right]. \quad (3)$$

3 Results and Discussions

This section explores the implications of the new CDF measurement on the 2HDM type-I. A systematic parameter space scan was performed using the code 2HDMC-1.8.0 [8], assuming the CP-even Higgs boson H aligns with the observed SM-like Higgs at $M_H = 125.09\,\text{GeV}$ [9].

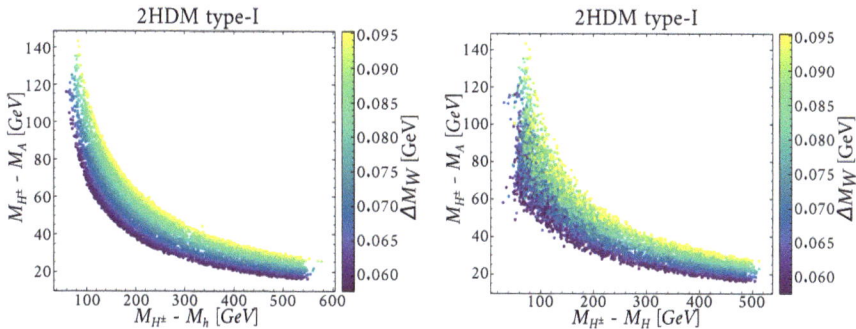

Fig. 1 Splitting of $M_{H^\pm} - M_A$ as a function of $M_{H^\pm} - M_h$ (left) and $M_{H^\pm} - M_H$ (right)

Theoretical and experimental constraints, including constraints from flavor physics, were considered during the scan, as detailed in Ref. [10]. The allowed parameter space was then further refined based on the $\chi^2_{M_W^{CDF}}$ test, considering only points within 2σ of the new CDF M_W measurement.

The dominant contribution to the W boson mass and $\sin^2\theta_{\text{eff}}$ observables is that of the T parameter. The new CDF measurement of the W boson mass necessitates the charged Higgs boson ($H^\pm$) to be heavier than the neutral Higgses (h, H, and A). This can be seen in Fig. 1, which illustrates the mass splitting $M_{H^\pm} - M_A$ as a function of $M_{H^\pm} - M_h$ (left panel) and $M_{H^\pm} - M_H$ (right panel), with the deviation from the SM prediction (ΔM_W) is color-coded. Scenarios where $M_{H^\pm}$ is lighter than the neutral Higgses are ruled out, as they lead to an insufficiently small or negative parameter T, incompatible with the CDF finding. Notably, the new CDF measurement significantly influences the charged Higgs mass, pushing it to be larger than 155 GeV.

4 Conclusion

In this paper, we addressed the recent deviation in the W boson mass measurement reported by CDF, showing a significant 7σ departure from the SM prediction. Our analysis in the 2HDM with an inverted hierarchy demonstrates that this discrepancy can be reconciled. We found that a positive T parameter, essential for aligning with the CDF findings, necessitates the charged Higgs boson mass, $M_{H^\pm}$, to be greater than M_h, M_H, and M_A. Cases with $M_{H^\pm}$ less than these masses fail to match the new CDF M_W measurement.

References

1. T. Aaltonen, et al., (CDF Collaboration), Science **376**, 170–176 (2022)
2. P.A. Zyla et al., Particle Data Group, Prog. Theor. Exp. Phys. **2020**(8), 083C01 (2020)
3. M. Awramik, et al., Phys. Rev. D **69**, 053006 (2004) [arXiv:hep-ph/0311148]
4. A. Broggio, et al., J. High Energy Phys. **11**, 058 (2014) [arXiv:1409.3199 [hep-ph]]
5. C.T. Lu, et al., [arXiv:2204.03796 [hep-ph]]
6. Y.Z. Fan, et al., [arXiv:2204.03693 [hep-ph]]
7. M.E. Peskin, T. Takeuchi, Phys. Rev. Lett. **65**, 964–967 (1990)
8. D. Eriksson, et al., Comput. Phys. Commun. **181**, 189–205 (2010) [arXiv:0902.0851]
9. G. Aad, et al., ATLAS and CMS Collaborations, Phys. Rev. Lett. **114**, 191803 (2015) [arXiv:1503.07589]
10. H. Abouabid, et al., Nucl. Phys. B **989**, 116143 (2023) [arXiv:2204.12018]

Open Access This chapter is licensed under the terms of the Creative Commons Attribution 4.0 International License (http://creativecommons.org/licenses/by/4.0/), which permits use, sharing, adaptation, distribution and reproduction in any medium or format, as long as you give appropriate credit to the original author(s) and the source, provide a link to the Creative Commons license and indicate if changes were made.

The images or other third party material in this chapter are included in the chapter's Creative Commons license, unless indicated otherwise in a credit line to the material. If material is not included in the chapter's Creative Commons license and your intended use is not permitted by statutory regulation or exceeds the permitted use, you will need to obtain permission directly from the copyright holder.

Propagation and Oscillations of Cosmic Neutrinos in a Stochastic Magnetic Field

Konstantin Kouzakov, Anastasia Nikolaeva, and Alexander Studenikin

1 Introduction

In theoretical extensions of the Standard Model massive neutrinos can have nonzero electromagnetic properties [1]. The most studied theoretically and searched experimentally among these properties is the neutrino magnetic moment. One of the manifestations of a nonzero neutrino magnetic moment can be spin oscillations of cosmic neutrinos propagating in an interstellar magnetic field [2]. The latter is known to have both deterministic and stochastic components. In this contribution we consider the neutrino evolution in a fluctuating interstellar magnetic field and outline how the stochastic magnetic-field component can impact the probabilities of flavor and spin oscillations of Dirac neutrinos in a two-component neutrino scenario.

2 Neutrino Evolution in a Fluctuating Magnetic Field

We consider the neutrino mass basis $(v_1^+, v_1^-, v_2^+, v_2^-)^T$, where the neutrino helicity basis states $|v_1^\pm\rangle$, $|v_2^\pm\rangle$ with masses m_1 and m_2 are related to the flavor ones as follows:

$$|v_e^{R,L}\rangle = \cos\theta |v_1^\pm\rangle + \sin\theta |v_2^\pm\rangle, \qquad |v_\mu^{R,L}\rangle = -\sin\theta |v_1^\pm\rangle + \cos\theta |v_2^\pm\rangle. \quad (1)$$

The study was supported by a grant from the Russian Science Foundation (project # 22-22-00384).

K. Kouzakov (✉) · A. Nikolaeva · A. Studenikin
Faculty of Physics, Lomonosov Moscow State University, Moscow, Russia
e-mail: kouzakov@gmail.com; n.nikolaeva11n.n11@gmail.com

© The Author(s) 2026
Y. Tayalati, M. Gouighri (eds.), *The First African Conference on High Energy Physics*, Springer Proceedings in Physics 425,
https://doi.org/10.1007/978-3-031-88933-2_32

The effective Hamiltonian of the problem is given by [2]

$$\hat{H}_{eff} = \hat{H}_{vac} + \hat{H}_B, \qquad (2)$$

where $\hat{H}_{vac} = \text{diag}(-\omega_\nu, -\omega_\nu, \omega_\nu, \omega_\nu)$ is the vacuum part, with $\omega_\nu = \frac{\Delta m^2}{4E_\nu}$, $\Delta m^2 = m_2^2 - m_1^2$, and E_ν being the neutrino energy. The Hamiltonian of the neutrino interaction with a magnetic field can be presented as $\hat{H}_B = B_\perp \hat{\mathcal{M}}$, where $B_\perp$ is the transverse magnetic-field component with respect to the neutrino velocity, and

$$\hat{\mathcal{M}} = \begin{pmatrix} 0 & \mu_1 & 0 & \mu_{12} \\ \mu_1 & 0 & \mu_{12} & 0 \\ 0 & \mu_{21} & 0 & \mu_2 \\ \mu_{21} & 0 & \mu_2 & 0 \end{pmatrix} \qquad (3)$$

is the neutrino magnetic moment matrix, with $\mu_{1(2)}$ ($\mu_{12(21)}$) being the diagonal (transition) neutrino magnetic moments. In the case of Dirac neutrinos the matrix $\hat{\mathcal{M}}$ is real and $\mu_{12} = \mu_{21}$, whereas for the Majorana neutrinos it is imaginary with $\mu_1 = \mu_2 = 0$ and $\mu_{12} = -\mu_{21} = -\mu_{12}^*$.

The galactic and extragalactic magnetic fields are composed of the large-scale regular component **B** and a small-scale stochastic component **h**. The latter results from interstellar fluctuations, galactic winds, cosmic turbulence and primordial magnetic field fluctuations and is characterized by the correlation function $\langle h_\alpha(t) h_\beta(0) \rangle = \frac{w^2}{2\mu_\nu^2} \delta(t)$, where μ_ν is a putative neutrino magnetic moment and $w^2 = 2\eta(\mu_\nu B)^2 L_0$ is the dissipation parameter, with L_0 being the correlation length. As a consequence, the density matrix of the system obeys the evolution equation in the form of the Lindblad master equation [3–5]:

$$\frac{d\hat{\varrho}}{dt} = -i\left[\hat{H}, \hat{\varrho}\right] - \frac{w^2}{4\mu_\nu^2}\left(\hat{\varrho}\hat{\mathcal{M}}^2 + \hat{\mathcal{M}}^2\hat{\varrho} - 2\hat{\mathcal{M}}\hat{\varrho}\hat{\mathcal{M}}\right). \qquad (4)$$

3 Neutrino Flavor and Spin Oscillation Probabilities

Considering Dirac neutrinos, we assume their transition magnetic moments to be zero, i.e., $\mu_{12} = \mu_{21} = 0$. For the initial neutrino state $\nu(0) = \nu_\mu^L$ the solution of the Lindblad master equation (4) yields the flavor-change probability as

$$P^D_{\nu_\mu^L \to \nu_e^L} = \frac{1}{4} \sin^2 2\theta \left\{ 1 + \frac{1}{2} \exp\left(-w^2 \frac{\mu_1^2}{\mu_\nu^2} t\right) \cos(2\omega_1 t) \right.$$
$$\left. + \frac{1}{2} \exp\left(-w^2 \frac{\mu_2^2}{\mu_\nu^2} t\right) \cos(2\omega_2 t) - \cos(2\omega_\nu t) \right\}$$

$$\times \left[\exp\left(-w^2 \frac{\mu_-^2}{\mu_\nu^2} t\right) \cos(2\omega_- t) + \exp\left(-w^2 \frac{\mu_+^2}{\mu_\nu^2} t\right) \cos(2\omega_+ t) \right] \Big\}, \tag{5}$$

where $\mu_\pm = (\mu_1 \pm \mu_2)/2$, $\omega_{1,2} = \mu_{1,2} B_\perp$ and $\omega_\pm = \mu_\pm B_\perp$.
The spin-flip probability $P^D_{\nu_\mu^L \to \nu^R}$ is determined by the expression:

$$P^D_{\nu_\mu^L \to \nu^R} = \frac{1}{2}\left[1 - \sin^2\theta \exp\left(-w^2 \frac{\mu_1^2}{\mu_\nu^2} t\right) \cos(2\omega_1 t)\right.$$
$$\left. - \cos^2\theta \exp\left(-w^2 \frac{\mu_2^2}{\mu_\nu^2} t\right) \cos(2\omega_2 t)\right]. \tag{6}$$

As opposed to the probability of the flavor oscillations (5), it does not depend on the neutrino energy E_ν. It can be seen from Eqs. (5) and (6) that the presence of a stochastic magnetic-field component leads to the loss of quantum coherence in neutrino oscillations, with the effective decoherence rate $\Gamma \sim w^2$.

4 Summary and Conclusions

The Lindblad master equation is solved for Dirac neutrinos with a nonzero magnetic moment propagating in a fluctuating magnetic field. The probabilities of neutrino flavor and spin oscillations are derived in a two-component framework. Our results can be used for studying the effect of interstellar magnetic fields on the propagation and oscillations of cosmic neutrinos. The indicated effect can influence the registration of ultrahigh-energy neutrinos in IceCube, Baikal-GVD and KM3NeT. It can also affect the supernova neutrino signals in upcoming experiments such as JUNO, Hyper-Kamiokande and DUNE.

Acknowledgments We are grateful to Konstantin Stankevich and Artem Popov for useful discussions. The study is conducted within the scientific program of the National Center for Physics and Mathematics (Section No. 8, Stage 2023–2025).

References

1. C. Giunti, A. Studenikin, Neutrino electromagnetic interactions: A window to new physics. Rev. Mod. Phys. **87**(2), 531–591 (2015). https://doi.org/10.1103/RevModPhys.87.531
2. P. Kurashvili, K.A. Kouzakov, L. Chotorlishvili, A.I. Studenikin, Spin-flavor oscillations of ultrahigh-energy cosmic neutrinos in interstellar space: the role of neutrino magnetic moments. Phys. Rev. D **96**(10), 103017 (2017). https://doi.org/10.1103/PhysRevD.96.103017

3. P. Kurashvili, L. Chotorlishvili, K.A. Kouzakov, A.G. Tevzadze, A.I. Studenikin, Quantum witness and invasiveness of cosmic neutrino measurements. Phys. Rev. D **103**(3), 036011 (2021). https://doi.org/10.1103/PhysRevD.103.036011
4. P. Kurashvili, L. Chotorlishvili, K.A. Kouzakov, A.I. Studenikin, Coherence and mixedness of neutrino oscillations in a magnetic field. Eur. Phys. J. C **81**(4), 323 (2021). https://doi.org/10.1140/epjc/s10052-021-09039-2
5. P. Kurashvili, L. Chotorlishvili, K.A. Kouzakov, A.I. Studenikin, Quantum spin-flavour memory of ultrahigh-energy neutrino. Eur. Phys. J. Plus **137**(2), 234 (2022). https://doi.org/10.1140/epjp/s13360-022-02457-5

Open Access This chapter is licensed under the terms of the Creative Commons Attribution 4.0 International License (http://creativecommons.org/licenses/by/4.0/), which permits use, sharing, adaptation, distribution and reproduction in any medium or format, as long as you give appropriate credit to the original author(s) and the source, provide a link to the Creative Commons license and indicate if changes were made.

The images or other third party material in this chapter are included in the chapter's Creative Commons license, unless indicated otherwise in a credit line to the material. If material is not included in the chapter's Creative Commons license and your intended use is not permitted by statutory regulation or exceeds the permitted use, you will need to obtain permission directly from the copyright holder.

Weak/Strong Duality and the Asymptotic Weak Gravity Conjecture

Mohammed Charkaoui, R. Sammani, E. H. Saidi, and R. Ahl Laamara

1 Introduction

The Asymptotic Weak Gravity Conjecture [1–3] has been proposed combining notions from the Weak Gravity Conjecture [4, 5] and the Distance Conjecture [6, 7]. It states that in an infinite distance in the moduli space, an infinite tower of states satisfies the Weak Gravity Conjecture (WGC). This large distance corresponds to weak coupling limits, which can be naively be seen as the limit $g \to 0$ where g is the gauge coupling.

The conjecture has been studied in M-theory on Calabi-Yau threefold [1] and fourfold [2, 3] with finite volume. In the original statement [1, 2], the conjecture focuses in the weak coupling limit. In [3] a particular geometrical structure given by $CY4 = K3 \times K3$ exhibiting an interesting duality has been presented relating weak and strong gauge couplings via linking the fiber and the base. The conjecture was then shown to hold for both weakly and strongly coupled states. In this paper we show that this duality can be applied to Calabi-Yau threefolds with finite volume, and thus we expect more towers given by M2 branes wrapping some 2-cycles which are BPS, or excitations of the string given by M5 on K3 as discussed in [3].

M. Charkaoui (✉) · R. Sammani · E. H. Saidi · R. Ahl Laamara
LPHE-MS, Science Faculty, Mohammed V University, Rabat, Morocco

Center of Physics and Mathematics CPM, Agdal Rabat, Morocco
e-mail: mohammed_charkaoui3@um5.ac.ma

2 Fibration Structure and Weak/Strong Gauge Duality

The study of the fibration structure of Calabi-Yau threefold with finite volume in an infinite distance of the Kahler moduli space has been studies in [1, 8]. It was shown that in an infinite distance in the moduli space a Calabi-Yau threefold with finite volume should have either a T^2 fibration or a $K3/T4$ fiber. These 2 cases were termed type T^2 and type $K3/T^4$.

The 2 Types were defined as follows:

- Type T^2: $\mathcal{J}_b^3 = 0$, $\mathcal{J}_b^2 \neq 0$
  To simplify we write the Kahler form as:

$$J = \lambda^{\frac{1}{3}} \mathcal{J}_b + \frac{1}{\lambda^{\frac{2}{3}}} \mathcal{J}_f \tag{1}$$

In such a way that the fiber T^2 contracts and the base expands $\mathcal{V}_{T^2} \sim \lambda^{-\frac{2}{3}}$, $\mathcal{V}_{\mathbb{B}_2} \sim \lambda^{\frac{2}{3}}$, while the overall volume remains finite.

- Type $K3/T^4$: $\mathcal{J}_b^3 = 0$, $\mathcal{J}_b^2 \neq 0$

$$J = \lambda^{\frac{2}{3}} \mathcal{J}_b + \frac{1}{\lambda^{\frac{1}{3}}} \mathcal{J}_f \tag{2}$$

and the volumes of the fiber and the base behave as $\mathcal{V}_{K3/T^4} \sim \lambda^{-\frac{2}{3}}$, $\mathcal{V}_{\mathbb{B}_1} \sim \lambda^{\frac{2}{3}}$.

For our case we will allow both the fiber and the base to either contract or expand as long as the volume of the internal manifold is finte. This implies other conditions, for the case of type T^2 the base should obey:

$$\mathcal{J}_f^3 = 0 \quad , \quad \mathcal{J}_f^2 \neq 0$$

which is the condition of a Calabi-Yau to admit a $K3$ or T^4 fibration. Similarly for the case of type $K3/T^4$ if the fiber expands the additional condition reads as:

$$\mathcal{J}_f^3 = 0 \quad , \quad \mathcal{J}_f^2 \neq 0 \tag{3}$$

Which is the condition of a Calabi-Yau admitting a T^2 fibration. Hence we see that (focusing on the case of K3) the Calabi-Yau threefold takes the form $CY3 = T^2 \times K3$.

Notice that the mapping $\lambda \to 1/\lambda$ exchanges (1) and (2) thus exchanging the roles of the fiber and the base. On the other hand in the limit $\lambda \to \infty$ where the fiber T^2 shrinks the gauge coupling matrix behaves as:

$$G_{i_f j_f} \sim \lambda^{\frac{2}{3}} \quad , \quad G_{\alpha_b \beta_b} \sim \frac{1}{\lambda^{\frac{1}{3}}} \tag{4}$$

clearly the operation $\lambda \to 1/\lambda$ switches strong and weak couplings between the fiber and the base.

The limit $\lambda \to \infty$: the fiber T^2 shrinks and we distinguish two towers of BPS states [1] that satisfy the conjecture, since BPS states in 5D are superextremal [9]. one tower is given by M2 branes wrapping fibral curves, and others wrapping curves with positive self-intersection in the base K3, as for non-non-BPS states it remains an open question if they are superextremal in this limit.

The limit $\lambda \to 0$: the fiber $K3$ shrinks, according to [1] two towers of states given the BPS states mentioned earlier and non-BPS states given by excitations of the heterotic string dual to M5/K3, moreover the base curves host also towers of BPS states, all these towers satisfy the conjecture, provided that the curves in K3 do not degenerate at an infinite distance in the moduli space according to [1].

3 Conclusion and Comments

In this paper we investigated the Asymptotic WGC in 5D coming from M-Theory on a Calabi-Yau threefold with finite volume. We have used results from [3] on the duality between the different gauge regimes, in addition to [1] on the fiber structure of the manifold as well as the states satisfying the WGC. We showed that by allowing both the fiber and the base to expand or shrink while keeping the volume of the manifold finite, we get a symmetry exchanging weak and strong gauge couplings. This will allow us to build more towers of states satisfying the WGC. These towers populate directions in the charge lattice dual strongly or weakly coupled gauge groups. This result can also be generalised to M-theory on Calabi-Yau fourfold for setups other than CY4 $= K3 \times K3$.

Following the classification of fiber structure of Calabi-Yau fourfolds in [3]. This results of this investigation can also be generalised to find a duality between what was labeled Type-T^2 and type-$\mathbb{V}$. Thus generating a symmetry between strong and weak gauge couplings and provide as a result more towers of states satisfying the conjecture.

References

1. C.F. Cota, A. Mininno, T. Weigand, M. Wiesner, The asymptotic weak gravity conjecture in M-theory. J. High Energy Phys. **2023**(8), 1–49 (2023)
2. C.F. Cota, A. Mininno, T. Weigand, M. Wiesner, The asymptotic weak gravity conjecture for open strings. J. High Energy Phys. **11**, 58 (2022). https://doi.org/10.1007/s13160-022-0158-0
3. M. Charkaoui, R. Sammani, E.H. Saidi, R.A. Laamara, Asymptotic weak gravity conjecture in M-theory on K 3× K 3. Prog. Theor. Exp. Phys. **2024**(7), 073B08 (2024)
4. N. Arkani-Hamed, L. Motl, A. Nicolis, C. Vafa, The string landscape, black holes, and gravity as the weakest force. J. High Energy Phys. **2007**(6), 60 (2007)

5. B. Heidenreich, M. Lotito, Proving the weak gravity conjecture in perturbative string theory, part I: the bosonic string (preprint, 2024). arXiv:2401.14449
6. H. Ooguri, C. Vafa, On the geometry of the string landscape and the swampland. Nucl. Phys. B **766**(1–3), 21–33 (2007)
7. A. Castellano, I. Ruiz, I. Valenzuela, Stringy evidence for a universal pattern at infinite distance (preprint, 2023). arXiv:2311.01536
8. S.J. Lee, W. Lerche, T. Weigand, Emergent strings from infinite distance limits. J. High Energy Phys. **2022**(2), 1–105 (2022)
9. M. Alim, B. Heidenreich, T. Rudelius, The weak gravity conjecture and BPS particles. Fortschr. Phys. **69**(11–12), 2100125 (2021)

Open Access This chapter is licensed under the terms of the Creative Commons Attribution 4.0 International License (http://creativecommons.org/licenses/by/4.0/), which permits use, sharing, adaptation, distribution and reproduction in any medium or format, as long as you give appropriate credit to the original author(s) and the source, provide a link to the Creative Commons license and indicate if changes were made.

The images or other third party material in this chapter are included in the chapter's Creative Commons license, unless indicated otherwise in a credit line to the material. If material is not included in the chapter's Creative Commons license and your intended use is not permitted by statutory regulation or exceeds the permitted use, you will need to obtain permission directly from the copyright holder.

Neutrino Oscillations and Quantum Decoherence

Konstantin Stankevich ⓘ, Alexander Studenikin ⓘ, and Maxim Vyalkov ⓘ

Neutrino quantum decoherence is the process of the violation of the coherent superposition of neutrino states engendered by the neutrino interaction with external environment. In [1–3] we proposed and considered the new mechanism of the neutrino quantum decoherence engendered by neutrino decay to a lighter neutrino and an arbitrary massless particle. In this paper we consider that there are two possible independent neutrino decays: (1) neutrino decay to a lighter state and a massless particle and (2) neutrino decay to a lighter state and another massless particle. Photons, dark photons, axions, etc. can be considered as an example of these massless particles.

The evolution of the entire system of both neutrinos and external environment (reservoir consisted of two types of massless particles) can be represented as follows

$$\rho(t) = U(t)\, \rho_0\, U^\dagger(t), \tag{1}$$

where $\rho(t)$ and $U(t)$ is the density matrix and evolution operator of the entire system, ρ_0 is the density matrix of the system at the initial moment of time. According to the approach proposed in [1–3] we use an expression for the expansion of the evolution operator of the entire system up to the second order in powers of the coupling constant

---

K. Stankevich · A. Studenikin
Faculty of Physics, Lomonosov Moscow State University, Moscow, Russia
e-mail: studenik@srd.sinp.msu.ru

M. Vyalkov (✉)
Faculty of Physics, Lomonosov Moscow State University, Moscow, Russia

Branch of Lomonosov Moscow State University in Sarov, Nizhny Novgorod, Russia
e-mail: vyalkovmsu@yandex.ru

$$U(t, t_0) = 1 + (-i) \int_{t_0}^{t} dt_1 H_I(t_1) + (-i)^2 \int_{t_0}^{t} dt_1 \int_{t_0}^{t_1} dt_2 H_I(t_1) H_I(t_2) + \ldots \tag{2}$$

where $H_I(t)$ is the Hamiltonian of the interaction of the neutrino with external reservoir. To get an expression for the neutrino density matrix, we take the trace of the external environment states of freedom $\rho_v(t) = \text{Tr}_A \rho(t)$. Finally, for the neutrino density matrix we get

$$\rho(t) = \rho_0 + i \int_{t_0}^{t} [H_I, \rho] dt +$$
$$+ \frac{1}{2} \bigg(\int_{t_0}^{t} \int_{t_0}^{t} dt_1 dt_2 H_I(t_2) \rho H_I(t_1) + \int_{t_0}^{t} \int_{t_0}^{t} dt_1 dt_2 H_I(t_1) \rho H_I(t_2) -$$
$$- \int_{t_0}^{t} dt_1 \int_{t_0}^{t} dt_2 \underset{\rightarrow}{T} \{H_I(t_1) H_I(t_2)\} \rho$$
$$- \rho \int_{t_0}^{t} dt_1 \int_{t_0}^{t} dt_2 \underset{\leftarrow}{T} \{H_I(t_1) H_I(t_2)\} \bigg). \tag{3}$$

The neutrino interaction with an external environment is described by two independent interaction Hamiltonians

$$H_I(t_2) = H_1(t_2) + H_2(t_2). \tag{4}$$

Following the approach [1–3] we get the expression for the neutrino evolution

$$\frac{d\rho_v(t)}{dt} = -i [H_1(x) + H_2(x), \rho_v(t)] + D[\rho_v(t)], \tag{5}$$

where the term $D[\rho_v(t)]$ is responsible for neutrino decoherence engendered by neutrino decays. The neutrino decoherence parameter Γ that describes the amplitude of the effect turned out to be a sum of the two independent neutrino decoherence parameters Γ_1 and Γ_2 engendered by the first channel of the neutrino decay and the second channel correspondingly, i.e. $\Gamma = \Gamma_1 + \Gamma_2$. In other words there is no interference terms in neutrino quantum decoherence when considering two independent channels of the neutrino decay.

Acknowledgments One of the authors (A.S.) is thankful to the organizers of the First African Conference on High Energy Physics for the hospitality. This study was supported by the Russian Science Foundation (project No. 22-22-00384).

This study also Supported by the Scientific Program of the National Center for Physics and Mathematics, Section No. 8 (Stage 2023-2025).

References

1. K. Stankevich, A. Studenikin, Neutrino quantum decoherence engendered by neutrino radiative decay. Phys. Rev. D **101**(5), 056004 (2020). https://doi.org/10.1103/PhysRevD.101.056004
2. A. Lichkunov, K. Stankevich, A. Studenikin, M. Vyalkov, Neutrino quantum decoherence engendered by neutrino decay to photons, familons, and gravitons. J. Phys. Conf. Ser. **2156**(1), 012240 (2021). https://doi.org/10.1088/1742-6596/2156/1/012240
3. A. Lichkunov, K. Stankevich, A. Studenikin, M. Vyalkov, Neutrino decay processes and flavour oscillations. PoS EPS-HEP2021 202 (2022). https://doi.org/10.22323/1.398.0202

Open Access This chapter is licensed under the terms of the Creative Commons Attribution 4.0 International License (http://creativecommons.org/licenses/by/4.0/), which permits use, sharing, adaptation, distribution and reproduction in any medium or format, as long as you give appropriate credit to the original author(s) and the source, provide a link to the Creative Commons license and indicate if changes were made.

The images or other third party material in this chapter are included in the chapter's Creative Commons license, unless indicated otherwise in a credit line to the material. If material is not included in the chapter's Creative Commons license and your intended use is not permitted by statutory regulation or exceeds the permitted use, you will need to obtain permission directly from the copyright holder.

Neutrino Oscillations in External Environment

A. Popov, N. Dolganov, V. Shakhov, Konstantin Stankevich ⓘ, and A. Studenikin ⓘ

1 Neutrino Effective Hamiltonian

Consider neutrino flavour, spin and spin-flavour oscillations engendered by neutrino interactions with an external electric current due to neutrino charge radii and anapole moments for the case of two flavour neutrinos with two possible helicities $v_f = \left(v_e^-, v_\alpha^-, v_e^+, v_\alpha^+\right)$, where $\alpha = \mu$ or τ. The scheme of the calculations performed is described in [1, 2].

In the mass basis the neutrino effective potential describing electromagnetic interactions of the neutrino field v with the external electric current is given by

$$H_J^{(m)fi} = \lim_{q \to 0} \frac{1}{T} \frac{\langle v_f(p_f, h_f) | \int d^4x \mathcal{H}_J | v_i(p_i, h_i) \rangle}{\langle v(p, h) | v(p, h) \rangle}, \qquad (1)$$

where $q = p_i - p_f$, p_i and p_f are the initial and final neutrino momenta, T is the normalization time. The matrix element

$$\langle v_f(p_f, h_f) | \mathcal{H}_J | v_i(p_i, h_i) \rangle = \bar{u}_f(p_f, h_f) \Lambda_\mu^{fi}(q) \frac{1}{q^2} u_i(p_i, h_i) J_{EM}^\mu e^{-iqx}. \qquad (2)$$

is determined by the electric current of the external charged fermions f (the protons or electrons), $J_{EM}^\mu = e(n_f, n_f \boldsymbol{v}_f)$, n_f is density of fermions f. Note that Λ_μ^{fi} in (2) contains the corresponding terms of the neutrino electromagnetic vertex (for its decomposition see [3]). If one takes into account only the electric charge and anapole form factors (which is the subject of this paper) then

---

A. Popov (✉) · N. Dolganov · V. Shakhov · K. Stankevich · A. Studenikin
Faculty of Physics, Lomonosov Moscow State University, Moscow, Russia
e-mail: ar.popov@physics.msu.ru; kl.stankevich@physics.msu.ru

$$\Lambda_\mu^{fi}(q) = (\gamma_\mu - q_\mu \gamma_\nu q^\nu/q^2)\left[f_Q^{fi}(q^2) + f_A^{fi}(q^2)q^2\gamma_5\right], \qquad (3)$$

where $f_Q^{fi}(q^2)$ and $f_A^{fi}(q^2)$ are charge and anapole form factors in the neutrino mass basis. The neutrino charge radius is determined by the second term in the expansion of the neutrino charge form factor $f_Q(q^2) = f_Q(0) + q^2 \left.\frac{df_Q(q^2)}{dq^2}\right|_{q^2=0}$ and the charge radius is given by $\langle r^2 \rangle = 6 \left.\frac{df_Q(q^2)}{dq^2}\right|_{q^2=0}$. We consider the case of zero neutrino millicharge $f_Q(0) = 0$. Therefore, the electromagnetic vertex accounted for the charge radius $\langle r^2 \rangle$ and the anapole moment $a = f_A(0)$ reads

$$\Lambda_\mu^{fi}(q) = (q^2\gamma_\mu - q_\mu \gamma_\nu q^\nu)\left[\frac{\langle r^2\rangle^{fi}}{6} + a^{fi}\gamma_5\right] \qquad (4)$$

which leads to the following expression for the effective interaction Hamiltonian:

$$H_J^{(m)fi} = 2\chi^{(h_f)\dagger}\left\{J_\parallel^{EM}\left(\frac{\langle r^2\rangle^{fi}}{6} + a^{fi}\sigma_3\right)\right.$$
$$+ J_\perp^{EM}\left[\left(\sigma_1\gamma_{fi}^{-1}\cos\xi + \sigma_2\gamma_{fi}^{-1}\sin\xi\right)a^{fi} +\right.$$
$$\left.\left. + \left(i\sigma_1\tilde\gamma_{fi}^{-1}\sin\xi - i\sigma_2\tilde\gamma_{fi}^{-1}\cos\xi\right)\frac{\langle r^2\rangle^{fi}}{6}\right]\right\}\chi^{(h_i)}, \qquad (5)$$

where orthogonal xyz-coordinate system was introduced, neutrinos propagate along z-axis, $J_\parallel^{EM}$ and $J_\perp^{EM}$ are the longitudinal and transversal (in respect to z-axis) electric currents, ξ is the angle between the x-axis and $J_\perp^{EM}$, $\chi^{(h)}$ defines the neutrino helicity states $\chi^{(+)} = (1, 0)^T$ and $\chi^{(-)} = (0, 1)^T$. The gamma factors are given by: $\gamma_\alpha^{-1} = \frac{m_\alpha}{E_\alpha}$, $\gamma_{\alpha\beta}^{-1} = \frac{1}{2}\left(\gamma_\alpha^{-1} + \gamma_\beta^{-1}\right)$, $\tilde\gamma_{\alpha\beta}^{-1} = \frac{1}{2}\left(\gamma_\alpha^{-1} - \gamma_\beta^{-1}\right)$.

Now consider neutrino spin oscillations $\nu_e^L \leftrightarrow \nu_e^R$ accounting for neutrino weak interaction with matter, electromagnetic interactions with an external electric current due to the neutrino charge radius and anapole moment and electromagnetic interaction with an external magnetic field due to the neutrino magnetic moment.

2 Neutrino Spin Oscillations $\nu_e^L \leftrightarrow \nu_e^R$

Consider two neutrino states with different helicities: (ν_e^L, ν_e^R). The corresponding neutrino oscillations are governed by the evolution equation

$$i\frac{d}{dt}\begin{pmatrix}v_e^L\\v_e^R\end{pmatrix} = H\begin{pmatrix}v_e^L\\v_e^R\end{pmatrix}$$ with Hamiltonian (for the first two terms see [1])

$$H = \frac{G_F}{2\sqrt{2}}\begin{pmatrix}4n_e(1-v_{e\|}) - 2n_n(1-v_{n\|}) & (2n_e v_{e\perp} - n_n v_{n\perp})\left(\frac{\eta}{\gamma}\right)_{ee}\\(2n_e v_{e\perp} - n_n v_{n\perp})\left(\frac{\eta}{\gamma}\right)_{ee} & 0\end{pmatrix} +$$

$$\begin{pmatrix}\left(\frac{\mu}{\gamma}\right)_{ee}B_\| & -\mu_{ee}B_\perp e^{i\phi}\\-\mu_{ee}B_\perp e^{-i\phi} & -\left(\frac{\mu}{\gamma}\right)_{ee}B_\|\end{pmatrix} + \begin{pmatrix}-2J_\|^{EM}a^{ee} & 2J_\perp^{EM}\left(\frac{a}{\gamma}\right)_{ee}e^{i\xi}\\2J_\perp^{EM}\left(\frac{a}{\gamma}\right)_{ee}e^{-i\xi} & 2J_\|^{EM}a^{ee}\end{pmatrix},$$
(6)

where n_n and n_e are the neutron and electron densities, $\mathbf{v}_{n,e} = \mathbf{v}_{n,e\|} + \mathbf{v}_{n,e\perp}$ are the neutrons and electrons velocities, $\mathbf{B} = \mathbf{B}_\| + \mathbf{B}_\perp$ is an external magnetic field, ϕ is the angle between $\mathbf{v}_\perp$ and $\mathbf{B}_\perp$, $\mu_{ee} = \mu_{11}\cos^2\theta + \mu_{22}\sin^2\theta + \mu_{12}\sin 2\theta$, $\mu_{ee} = \mu_{11}\cos^2\theta + \mu_{22}\sin^2\theta + \mu_{12}\sin 2\theta$,

$$\left(\frac{\eta}{\gamma}\right)_{ee} = \frac{\cos^2\theta}{\gamma_{11}} + \frac{\sin^2\theta}{\gamma_{22}}, \quad \left(\frac{\mu}{\gamma}\right)_{ee} = \frac{\mu_{11}}{\gamma_{11}}\cos^2\theta + \frac{\mu_{22}}{\gamma_{22}}\sin^2\theta + \frac{\mu_{12}}{\gamma_{12}}\sin 2\theta,$$

$$\left(\frac{a}{\gamma}\right)_{ee} = \frac{a^{11}}{\gamma_{11}}\cos^2\theta + \frac{a^{22}}{\gamma_{22}}\sin^2\theta + \frac{a^{12}}{\gamma_{12}}\sin 2\theta.$$

For the oscillation $v_e^L \leftrightarrow v_e^R$ probability we get

$$P(x) = \sin^2 2\theta_{\text{eff}}\sin^2\frac{\pi x}{L_{\text{eff}}}, \quad \sin^2 2\theta_{\text{eff}} = \frac{E_{\text{eff}}^2}{E_{\text{eff}}^2 + \Delta_{\text{eff}}^2}, \quad L_{\text{eff}} = \frac{\pi}{\sqrt{E_{\text{eff}}^2 + \Delta_{\text{eff}}^2}},$$
(7)

where E_{eff}^2 and Δ_{eff}^2 are expressed in terms of the elements H_{ij} of the Hamiltonian (6):

$$E_{eff}^2 = 4|H_{12}|^2 = 4\left[\frac{G_F}{2\sqrt{2}}(2n_e v_{e\perp} - n_n v_{n\perp})\left(\frac{\eta}{\gamma}\right)_{ee} -\right.$$
$$\left. -\mu_{ee}B_\perp\cos\phi + 2J_\perp^{EM}\left(\frac{a}{\gamma}\right)_{ee}\cos\xi\right]^2 + 4\left[\mu_{ee}B_\perp\sin\phi - 2J_\perp^{EM}\left(\frac{a}{\gamma}\right)_{ee}\sin\xi\right]^2,$$
(8)

$$\Delta_{eff}^2 = (H_{11} - H_{22})^2 =$$
$$= \left[\frac{G_F}{\sqrt{2}}\left(2n_e(1-v_{e\|}) - n_n(1-v_{n\|})\right) + 2\left(\frac{\mu}{\gamma}\right)_{ee}B_\| - 4J_\|^{EM}a^{ee}\right]^2.$$
(9)

It follows that whereas the spin oscillations can be generated by the neutrino anapole moment interactions with an external electric current, the interaction due to the charge radius does not produce the spin oscillations. Thus, these peculiarities can be used for disintegration of the anapole moment and charge radius effects in neutrino oscillations.

Acknowledgments This study was supported by the Russian Science Foundation (project no. 24-12-00084). The work was performed using the scientific infrastructure provided within the scientific program of the National Center for Physics and Mathematics (section no. 8 "Physics of hydrogen isotopes", project "Fundamental studies in the field of neutrino physics and neutron-rich nuclei using isotopes of hydrogen and helium").

References

1. P. Pustoshny, A. Studenikin, Neutrino spin and spin-flavour oscillations in transversal matter currents with standard and non-standard interactions. Phys. Rev. D **98**(11), 113009 (2018). https://doi.org/10.1103/PhysRevD.98.113009
2. V. Shakhov, K. Stankevich, A. Studenikin, Spin and spin-flavour oscillations due to neutrino charge radii interaction with an external environment. J. Phys. Conf. Ser. **2156**, 012241 (2021). https://doi.org/10.1088/1742-6596/2156/1/012241
3. C. Giunti, A. Studenikin, Neutrino electromagnetic interactions: a window to new physics. Rev. Mod. Phys. **87**, 531 (2015). https://doi.org/10.1103/RevModPhys.87.531

Open Access This chapter is licensed under the terms of the Creative Commons Attribution 4.0 International License (http://creativecommons.org/licenses/by/4.0/), which permits use, sharing, adaptation, distribution and reproduction in any medium or format, as long as you give appropriate credit to the original author(s) and the source, provide a link to the Creative Commons license and indicate if changes were made.

The images or other third party material in this chapter are included in the chapter's Creative Commons license, unless indicated otherwise in a credit line to the material. If material is not included in the chapter's Creative Commons license and your intended use is not permitted by statutory regulation or exceeds the permitted use, you will need to obtain permission directly from the copyright holder.

Effect of Quantum Decoherence on Collective Neutrino Oscillations

Artem Popov, Anastasiia Purtova ⓘ, Konstantin Stankevich ⓘ, and Alexander Studenikin ⓘ

We describe the evolution of the neutrino and antineutrino in flavor basis in terms of the density matrices $\rho(t)$ and $\bar{\rho}(t)$, respectively, using the Lindblad equation:

$$\frac{d\rho(t)}{dt} = -i[H, \rho(t)] + D[\rho(t)]. \tag{1}$$

The similar equation hold for the antineutrino evolution. The total Hamiltonian of the neutrino H includes three components $H = H_{vac} + H_\lambda + H_{\nu\nu}$. The vacuum term, H_{vac}, for a neutrino mode of energy E in the flavor basis using the neutrino mixing matrix U reads

$$H_{vac} = \frac{UM^2U^\dagger}{2E}, \quad M^2 := \mathrm{diag}(0, \Delta m_{21}^2, \Delta m_{31}^2), \quad \Delta m_{kl} := m_k^2 - m_l^2. \tag{2}$$

The second term, H_λ, describes the matter effects and is dominated by the charged current interaction of electron neutrinos with electrons of net density N_e. In the weak eigenstate basis it can be written as $H_\lambda = \sqrt{2}G_F \, \mathrm{diag}(N_e, 0, 0)$, where G_F is the Fermi coupling constant. As neutrinos are considered to be ultra-relativistic their four velocity vector is $v^\mu = (1, \mathbf{v})$ and $|\mathbf{v}| = 1$. Then, the interaction between a neutrino and other neutrinos of N discrete modes with velocities $\mathbf{v}_j$ reads

$$H_{\nu\nu} := \sqrt{2}G_F n_\nu \sum_{j=1}^{N}(1 - \mathbf{v}\mathbf{v}_j)\rho. \tag{3}$$

A. Popov (✉) · A. Purtova · K. Stankevich · A. Studenikin
Faculty of Physics, Moscow State University, Moscow, Russia
e-mail: ar.popov@physics.msu.ru; kl.stankevich@physics.msu.ru

This part of the Hamiltonian is proportional to the effective neutrino density $n_\nu := \frac{1}{2}(n_{\nu_e} - n_{\bar{\nu}_e} + n_{\nu_x} - n_{\bar{\nu}_x} + n_{\nu_y} - n_{\bar{\nu}_y})$ and the relative angle of the propagation direction of the neutrino modes.

The dissipator, $D[\rho]$, describes the quantum decoherence of neutrino states and is given by the expression

$$D[\rho] = \frac{1}{2} \sum_{k=1}^{8} \left[V_k, \rho V_k^\dagger \right] + \left[V_k \rho, V_k^\dagger \right], \qquad (4)$$

where V_k are the dissipative operators corresponding to the interaction of the neutrino with the reservoir.

Following the approach that is widely used in studies of collective neutrino oscillations to simplify the equations (e.g. [1]) we work in a basis spanned by $|\nu_e\rangle$, $|\nu_x\rangle$ and $|\nu_y\rangle$:

$$\begin{pmatrix} \nu_e \\ \nu_\mu \\ \nu_\tau \end{pmatrix} = R_{23}^\dagger \begin{pmatrix} \nu_e \\ \nu_x \\ \nu_y \end{pmatrix}, \quad R_{23} = \begin{pmatrix} 1 & 0 & 0 \\ 0 & c_{23} & s_{23} \\ 0 & -s_{23} & c_{23} \end{pmatrix}. \qquad (5)$$

Here and in the following we use the shorthand $c_{kl} := \cos(\theta_{kl})$, $s_{kl} := \sin(\theta_{kl})$, $C_{kl} := \cos(2\theta_{kl})$, $S_{kl} := \sin(2\theta_{kl})$ for a mixing angle θ_{kl}. For the mixing angles we have the following values [2]: $\theta_{12} = 33.62°$, $\theta_{23} = 47.2°$, $\theta_{13} = 8.54°$.

To study the instabilities of our system we can treat flavor correlations (off-diagonal terms of the density matrix) as plane waves $\rho_{ij} = Q\, e^{-i\Omega t}$. This is the basis of the linearized stability analysis. If the frequency Ω has a non-zero imaginary part, then flavor conversion occurs with exponentially growing factor.

We study a simple model of two colliding neutrino beams (one dimensional motion) and assume the following: antineutrinos are moving to the left (index l) and neutrinos are moving to the right (index r). These two beams denoted by

$$Q^l = \begin{pmatrix} \bar{\rho}_{ex} \\ \bar{\rho}_{ey} \\ \bar{\rho}_{xy} \end{pmatrix} \text{ and } Q^r = \begin{pmatrix} \rho_{ex} \\ \rho_{ey} \\ \rho_{xy} \end{pmatrix}. \qquad (6)$$

The elements are spanned by the off-diagonal elements of the neutrino and antineutrino density matrices, ρ and $\bar{\rho}$. The two beams are collected in a six dimensional vector $Q = (Q^r, Q^l)^T$. The density of neutrinos and antineutrinos is considered to be equal and independent of time and space, so the eigenvalue equation reduces to [3]

$$\left(\Omega + i \begin{pmatrix} \Gamma & 0 \\ 0 & \Gamma \end{pmatrix} \right) Q = \begin{pmatrix} |\omega|A_0 + |\omega|\eta B_0 + (\lambda - 2\mu)\Lambda & 2\mu\Lambda \\ -2\mu\Lambda & -(|\omega|A_0 + |\omega|\eta B_0 - (\lambda + 2\mu)\Lambda) \end{pmatrix} Q. \qquad (7)$$

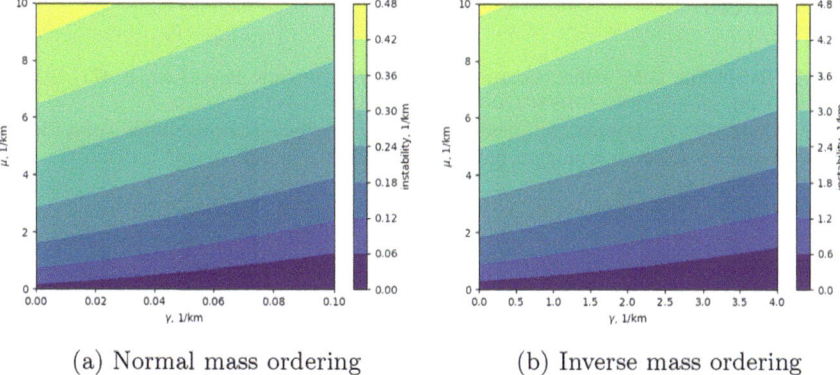

(a) Normal mass ordering   (b) Inverse mass ordering

Fig. 1 Dependence of the instabilities $Im(\Omega) - \gamma$ on the neutrino-neutrino interaction strength μ and decoherence parameter γ for both mass orderings. (**a**) Normal mass ordering. (**b**) Inverse mass ordering

Terms in the Hamiltonian are evaluated in accordance with [2]. Thus, the matrices A_0 and B_0 contain combinations of mixing angles and come from the vacuum term (2) and read

$$A_0 = \begin{pmatrix} -c_{12}^2 + c_{13}^2 s_{12}^2 & \frac{1}{2} S_{12} s_{13} & 0 \\ \frac{1}{2} S_{12} s_{13} & C_{13} s_{12}^2 & \frac{1}{2} S_{12} c_{13} \\ 0 & \frac{1}{2} S_{12} c_{13} & c_{12}^2 - s_{12}^2 s_{13}^2 \end{pmatrix}, \quad B_0 = \begin{pmatrix} s_{13}^2 & 0 & 0 \\ 0 & C_{13} & 0 \\ 0 & 0 & -c_{13}^2 \end{pmatrix}. \quad (8)$$

The matrix Λ is defined as $\Lambda = \text{diag}(1, 1, 0)$. And Γ is the decoherence matrix which in case of three neutrino flavors reads $\Gamma = \text{diag}(\gamma_{21}, \gamma_{31}, \gamma_{32})$. For simplicity in our qualitative analysis we consider $\gamma_{21} = \gamma_{31} = \gamma_{32} = \gamma$.

Equation (7) solved numerically for Ω, variable parameters are the strength of the neutrino-neutrino interaction μ and the decoherence parameter γ. Results are shown both for IO and NO on Fig. 1. From the obtained results we can see that in the case of a normal hierarchy the disappearance of the neutrino collective oscillations is possible for smaller values of the decoherence parameter (of the order of 0.1 km^{-1}).

Acknowledgments This study was supported by the Russian Science Foundation (project no. 24-12-00084). The work was performed using the scientific infrastructure provided within the scientific program of the National Center for Physics and Mathematics (section no. 8 "Physics of hydrogen isotopes", project "Fundamental studies in the field of neutrino physics and neutron-rich nuclei using isotopes of hydrogen and helium").

References

1. A.B. Balantekin, G.M. Fuller, Constraints on neutrino mixing. Phys. Lett. B **471**, 195–201 (1999). https://doi.org/10.1016/S0370-2693(99)01371-4
2. C. Döring, R.S.L. Hansen, M. Lindner, Stability of three neutrino flavor conversion in supernovae. J. Cosmol. Astropart. Phys. **08**, 003 (2019). https://doi.org/10.1088/1475-7516/2019/08/003
3. A.A. Purtova, K.L. Stankevich, A.I. Studenikin, Effect of quantum decoherence on collective neutrino oscillations. JETP Lett. **118**(2), 83–86 (2023). https://doi.org/10.1134/S0021364023601951

Open Access This chapter is licensed under the terms of the Creative Commons Attribution 4.0 International License (http://creativecommons.org/licenses/by/4.0/), which permits use, sharing, adaptation, distribution and reproduction in any medium or format, as long as you give appropriate credit to the original author(s) and the source, provide a link to the Creative Commons license and indicate if changes were made.

The images or other third party material in this chapter are included in the chapter's Creative Commons license, unless indicated otherwise in a credit line to the material. If material is not included in the chapter's Creative Commons license and your intended use is not permitted by statutory regulation or exceeds the permitted use, you will need to obtain permission directly from the copyright holder.

Hyper Kamiokande Energy Calibration with N16

Abderrazaq EL Abassi, Rafik Er-Rabit, and Mohamed Gouighri

1 The Hyper Kamiokande Experiment

Hyper Kamiokande (HK) [1] is a next generation neutrino experiment being built in Japan, more precisely, the far detector (FD) is under construction in Hida, Gifu, while other near detector (ND) facilities in Tokai, Ibaraki are being improved to meet the physics goals of the HK experiment. Led by the University of Tokyo and the High Energy Accelerator Research Organization (KEK), institutes from over 20 countries are collaborating to build the biggest detector of its kind in the world.

HK will follow the paths of its predecessors Super Kamiokande (SK) [2] and T2K experiments [3] in investigating the proton decay phenomenon, predicted by models beyond the Standard Model (BSM), and studying neutrinos from different sources, such us the cosmos, the Sun, the atmosphere, and the Earth, as well as exploring neutrino oscillations in lab-made accelerator neutrino and antineutrino beams. Data-taking is foreseen to start in 2027.

The underground water tank (c.f. Fig. 1) will be filled with 200 kilotons of ultra-pure water and will consist of an Inner Detector (ID) surrounded by an Outer Detector (OD). The ID walls will be equipped with 20,000 box-and-line (B&L) PMTs with a diameter of 50 cm manufactured by Hamamatsu, in addition to 800 multi-PMT modules (mPMTs), each consists of 19 units of 7.6 cm Hamamatsu PMTs, which will improve HK's vertexing and timing capabilities. The OD, which will be used as a veto shield from cosmic muons and other background sources, will be equipped with 3600 7.6 cm PMTs.

A. EL Abassi (✉) · R. Er-Rabit · M. Gouighri
Ibn Tofail University, Kenitra, Morocco
e-mail: abderrazaq.elabassi@uit.ac.ma; rafik.er-rabit@uit.ac.ma; mohamed.gouighri@uit.ac.ma

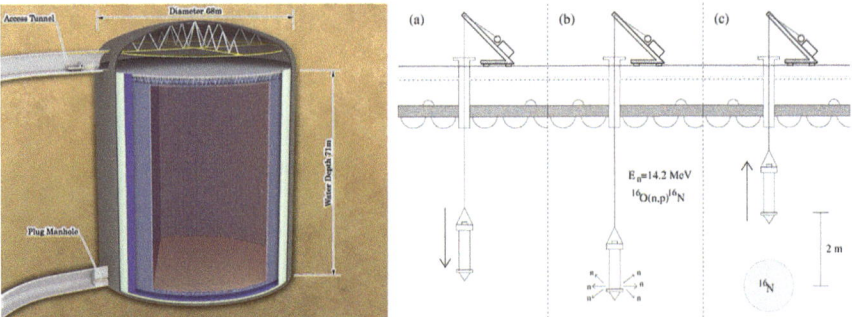

Fig. 1 (left) Hyper-Kamiokande far detector. (right) Calibration source deployment process

2 Water Activation Preliminary Results

The N16 cloud will be created in multiple places inside the water tank using a compact neutron generator, through the activation of O16 atoms in water. The N16 cloud will beta decay isotropically, and before starting to take data, the device will be raised above the radioactive cloud to minimize the shadowing effect [4].

In the following, we will present some preliminary results related to the water activation simulations and the generation of the N16 radioactive cloud, where we calculated the yield of N16 creation as a function of the incident neutron energies. The properties of the N16 cloud such as the spatial distribution have been presented.

Five million neutrons have been simulated using GEANT4 [5], where the neutron high precision package (NeutronHP) was used, the 14.2 MeV neutron beams are directed downwards matching the deployment orientation of the calibration devices inside the water tank, and an opening angle of 15 degrees was chosen (Fig. 2) which will be adapted to the real value once the properties of the deuterium tritium neutron generator (DTG) are known. For the given energy, a yield of 2.3% was obtained, higher than the 1.3% yield cited for the same energy value in the SK results 1].

The evolution of the N16 yield as function of the incident neutrons energy was studied for 100 energy values between 10 and 20 MeV (c.f. Fig. 2). The goal is to maximize this value to get enough statistics during calibration without the need to deploy our device at the same spot multiple times. The neutron capture energy threshold is around 11 MeV, with the yield peaking at the energy 11.7 MeV, however, the corresponding peak is narrow, and depending on the energy distribution spread of the generated neutrons (property of the device) the second peak around 14 MeV might be a better choice. Higher energies are ignored to avoid the creation of other radioactive isotopes like O15.

The spatial distribution of generated N16 isotopes was evaluated (c.f. Fig. 3). This information is crucial for a good vertex and energy reconstructions. The distributions have the expected shapes, where the symmetry around the Z axe implies a normal distribution for X and Y coordinates, while the vertical deployment

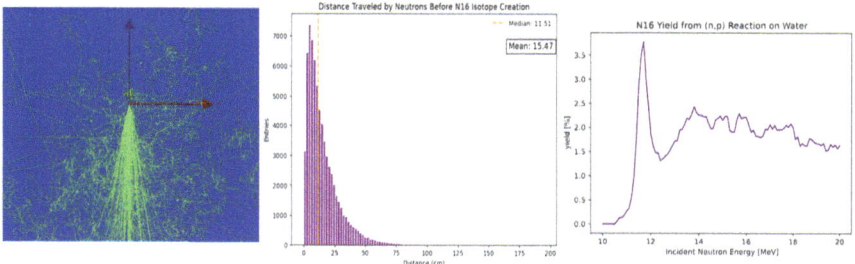

Fig. 2 (left) Visualization of the 200 neutron tracks. (middle) Distance traveled by neutrons before being captured. (right) N16 yield evolution as a function of incident neutron energies

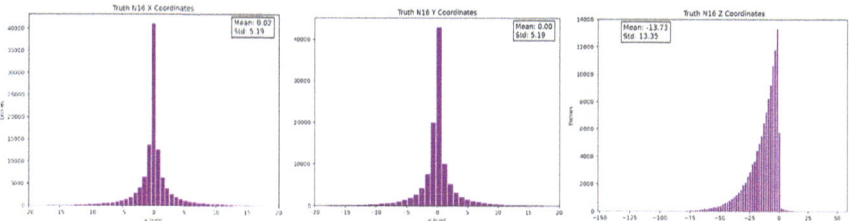

Fig. 3 Spatial distributions of the N16 isotopes for the deployment point (0, 0, 0)

of the source results in a skewed distribution for the Z coordinates. The average distance traveled by the neutrons before being captured is 15.47 cm, with half of them being captured within 11.51 cm from the deployment point (c.f. Fig. 2).

3 Conclusion

Future work includes the study of the N16 decay inside the water tank, the propagation of the daughter particles through the volume of the detector, and the reconstruction of the events, then the calculation of the energy resolution and the energy scale for multiple deployment points especially those close to the wall, to determine the optimal calibration strategy.

References

1. K. Abe et al. (Hyper-Kamiokande) (2018), 1805.04163.
2. S. Fukuda et al., The Super-Kamiokande detector. Nucl. Instrum. Methods Phys. Res. Sect. A **501** (2003)
3. K. Abe et al., The T2K experiment. Nucl. Instrum. Methods Phys. Res. Sect. A **659** (2011)

4. E. Blaufuss, G. Guillian, et al., Nucl .Instrum. Meth. A **458**(3), 638–649 (2001)
5. S. Agostinelli, and others, Nucl. Instrum. Meth. A **506**, 250–303 (2003)

Open Access This chapter is licensed under the terms of the Creative Commons Attribution 4.0 International License (http://creativecommons.org/licenses/by/4.0/), which permits use, sharing, adaptation, distribution and reproduction in any medium or format, as long as you give appropriate credit to the original author(s) and the source, provide a link to the Creative Commons license and indicate if changes were made.

The images or other third party material in this chapter are included in the chapter's Creative Commons license, unless indicated otherwise in a credit line to the material. If material is not included in the chapter's Creative Commons license and your intended use is not permitted by statutory regulation or exceeds the permitted use, you will need to obtain permission directly from the copyright holder.

Electromagnetic Effects in Deep Inelastic Neutrino-Proton Scattering

Konstantin Kouzakov, Elena Kovalevskaia, and Alexander Studenikin

1 Introduction

Deep inelastic scattering on nucleons is the main process of interaction of high-energy neutrinos with matter. The corresponding cross sections are measured, for example, in the IceCube experiment on the basis of the dependence of the neutrino flux registered in the range 60 TeV–10 PeV on the direction of its arrival in the detector [1]. This provides a unique opportunity to study the effects of physics beyond the Standard Model at high and ultrahigh neutrino energies. In particular, the search for the electromagnetic properties of neutrinos is of interest [2]. Indeed, the electromagnetic interactions of neutrinos are one of the fundamental problems in neutrino physics. They are studied theoretically and searched in experiments. The neutrino electromagnetic properties are related to the basics of elementary particle physics: since they are absent in the Standard Model with massless neutrinos, their discovery will open a window to new physics [2].

2 Neutrino Electromagnetic Form Factors

The electromagnetic properties of neutrinos are contained in the vertex function, which is a matrix in spinor space. It can be decomposed in terms of linearly independent products of Dirac matrices and the incoming and outgoing neutrino 4-momenta, k and k'.

In accordance with the Lorentz and gauge invariance, the electromagnetic vertex can be written as follows:

K. Kouzakov (✉) · E. Kovalevskaia · A. Studenikin
Faculty of Physics, Lomonosov Moscow State University, Moscow, Russia

© The Author(s) 2026
Y. Tayalati, M. Gouighri (eds.), *The First African Conference on High Energy Physics*, Springer Proceedings in Physics 425,
https://doi.org/10.1007/978-3-031-88933-2_38

$$\Lambda_\mu(q) = f_Q\left(q^2\right)\gamma_\mu - f_M\left(q^2\right)i\sigma_{\mu\nu}q^\nu + f_E\left(q^2\right)\sigma_{\mu\nu}q^\nu\gamma_5$$
$$+ f_A\left(q^2\right)\left(q^2\gamma_\mu - q_\mu \slashed{q}\right)\gamma_5, \qquad (1)$$

where $q = k - k'$ is the momentum transfer, $f_{Q,M,E,A}$ are the neutrino charge (Q), magnetic (M), electric (E), and anapole (A) form factors, which are Hermitian matrices in the space of neutrino mass states.

For brevity, we refer to the combination $f_Q - q^2 f_A$ as the *charge form factor*, denoting it by f_Q, and to the combination $f_M - if_E$ as the *magnetic form factor*, denoting it by f_M.

3 General Formalism

The neutral-current differential cross section for deep inelastic lepton-nucleon scattering in a laboratory frame is given by (see, for instance, [3])

$$\frac{d^2\sigma^{\ell N}}{dE'd\Omega} = \frac{\alpha^2}{m_N q^4}\frac{E'}{E}\sum_{j=\gamma,\gamma Z,Z}\eta_j L_j^{\mu\nu}W^j_{\mu\nu}, \qquad (2)$$

where E and E' are the initial and final lepton energies, and $L_j^{\nu\mu}$ and $W^j_{\nu\mu}$ are the leptonic and hadronic tensors, and

$$\eta_\gamma = 1, \qquad \eta_{\gamma Z} = \left(\frac{G_F M_Z^2}{2\sqrt{2}\pi\alpha}\right)\left(\frac{Q^2}{Q^2 + M_Z^2}\right), \qquad \eta_Z = \eta_{\gamma Z}^2.$$

The neutral-current total cross section derives from the expression

$$\sigma^{\ell N} = \int_{Q_0^2}^{s} dQ^2 \int_{Q^2/s}^{1} dx\, \frac{d^2\sigma^{\ell N}}{dx\,dQ^2}, \qquad (3)$$

with $Q_0 = 1$ GeV [4],

$$s = (k + P)^2 \approx m_N^2 + 2m_N E, \qquad Q^2 = -q^2 \approx 4EE'\sin^2\frac{\theta}{2}, \qquad x = \frac{Q^2}{2\nu m_N},$$

where k and P are the initial lepton and nucleon 4-momenta, θ is the lepton scattering angle, and

$$\nu = \frac{P\cdot q}{m_N} = E - E'.$$

4 Cross Sections

For the differential cross section of the $\nu_i p$-scattering, which accounts for electromagnetic (γ) and weak neutral-current (Z^0) interactions, we get

$$\frac{d^2\sigma^{\nu_i p}}{dx dQ^2} = \frac{4\pi\alpha^2}{xQ^4} \Bigg\{ \sum_f \left|f_Q^{fi}(Q^2)\right|^2 \left[xy^2 F_1^\gamma(x,Q^2) + \left(1-y-\frac{x^2y^2m_p^2}{Q^2}\right)\right.$$

$$\left. \times F_2^\gamma(x,Q^2)\right] - \eta_{\gamma Z} f_Q^{ii}(Q^2)\left[xy^2 F_1^{\gamma Z}(x,Q^2) + \left(1-y-\frac{x^2y^2m_p^2}{Q^2}\right)\right.$$

$$\left. \times F_2^{\gamma Z}(x,Q^2) \pm \left(y-\frac{y^2}{2}\right)xF_3^{\gamma Z}(x,Q^2)\right] + \eta_Z \Bigg[xy^2 F_1^Z(x,Q^2)$$

$$+ \left(1-y-\frac{x^2y^2m_p^2}{Q^2}\right) F_2^Z(x,Q^2) \pm \left(y-\frac{y^2}{2}\right)xF_3^Z(x,Q^2)\Bigg]$$

$$+ \frac{Q^2}{4xm_e^2} \sum_f \left|f_M^{fi}(Q^2)\right|^2 \left[-xy^2 F_1^\gamma(x,Q^2) + \frac{1}{2}(2-y)^2 F_2^\gamma(x,Q^2)\right] \Bigg\}, \quad (4)$$

where the $+ (-)$ sign refers to neutrinos (antineutrinos), and

$$y = \frac{P \cdot q}{P \cdot k} = \frac{\nu}{E}.$$

$f_{Q,M}^{fi}$ are the neutrino charge and magnetic form factors of diagonal ($i = f$) and transition ($i \neq f$) types given in units of elementary charge e_0 and Bohr magneton μ_B, respectively. $F_{1,2,3}^{\gamma,Z,\gamma Z}$ are dimensionless structure functions of the proton. Finally, the neutral-current total cross section is calculated on the basis of Eqs. (3) and (4).

Acknowledgments The study is conducted within the scientific program of the National Center for Physics and Mathematics (Section No. 8, Stage 2023-2025).

References

1. R. Abbasi, et al., (IceCube Collaboration), Measurement of the high-energy all-flavor neutrino-nucleon cross section with IceCube. Phys. Rev. D **104**(2), 022001 (2021). https://doi.org/10.1103/PhysRevD.104.022001
2. C. Giunti, A. Studenikin, Neutrino electromagnetic interactions: a window to new physics. Rev. Mod. Phys. **87**(2), 531–591 (2015). https://doi.org/10.1103/RevModPhys.87.531
3. E. Leader, E. Predazzi, *An Introduction to Gauge Theories and Modern Particle Physics* (Cambridge University Press, Cambridge, 1996). https://doi.org/10.1017/CBO9780511622595

4. C.A. Argüelles, F. Halzen, L. Wille, M. Kroll, M.H. Reno, High-energy behavior of photon, neutrino, and proton cross sections. Phys. Rev. D **92**(7), 074040 (2015). https://doi.org/10.1103/PhysRevD.92.074040

Open Access This chapter is licensed under the terms of the Creative Commons Attribution 4.0 International License (http://creativecommons.org/licenses/by/4.0/), which permits use, sharing, adaptation, distribution and reproduction in any medium or format, as long as you give appropriate credit to the original author(s) and the source, provide a link to the Creative Commons license and indicate if changes were made.

The images or other third party material in this chapter are included in the chapter's Creative Commons license, unless indicated otherwise in a credit line to the material. If material is not included in the chapter's Creative Commons license and your intended use is not permitted by statutory regulation or exceeds the permitted use, you will need to obtain permission directly from the copyright holder.

Dynamics of Quantum Resources in Hybrid System Under Decoherence Effect

Essalha Chaouki and Mostafa Mansour

1 Introduction

Quantum entanglement is a key resource for quantum information processing. However, it does not capture all non-classical correlations. Measures such as $\mathcal{U}_C$ and LQU have emerged to characterize these correlations. Practical applications are limited by decoherence, including ID, which can occur without external environmental interactions. Hybrid qubit-qutrit systems are known to be more resilient to such decoherence.

This study investigates the impact of ID and inhomogeneous magnetic fields on quantum correlations in a Heisenberg qubit-qutrit XXZ system with DM interaction. We assess skew information correlations and bipartite entanglement using LQU, $\mathcal{U}_C$, and $\mathcal{L}_\mathcal{N}$. Our results show that adjusting system parameters can mitigate the negative effects of decoherence.

The paper is structured as follows: Sect. 2 presents the quantifiers used; Sect. 3 details the system and the impact of decoherence; and Sect. 4 discusses our findings and concludes the study.

E. Chaouki (✉) · M. Mansour
LHEPMC, Department of Physics, Faculty of Sciences Aïn Chock, Hassan II University, Casablanca, Morocco

2 Quantum Correlation Measures

2.1 LQU

LQU ($\mathcal{U}$) is the minimum skew information in a quantum state due to the action of a local observable. For a state $\hat{D}$ with dimension $(2 \otimes d)$, $\mathcal{U}$ is given by

$$\mathcal{U}\left(\hat{D}\right) = 1 - \max\left(\Lambda_1, \Lambda_2, \Lambda_3\right) \tag{1}$$

where $\Lambda_{i=1,2,3}$ are the eigenvalues of the matrix $\mathcal{W}_{3\times 3}$, defined by:

$$\left(\mathcal{W}\right)_{ij} \equiv \text{Tr}\left\{\sqrt{\hat{D}}\left(\sigma_{Ai} \otimes \mathbb{I}_B\right)\sqrt{\hat{D}}\left(\sigma_{Aj} \otimes \mathbb{I}_B\right)\right\} \tag{2}$$

σ_{Ai} and $\mathbb{I}_B$ are the Pauli and identity operators acting on the subsystems A and B, respectively.

2.2 Uncertainty-Induced Nonlocality ($\mathcal{U}_C$)

$\mathcal{U}_C$ also stems from the Wigner–Yanase skew information (WYSI). For a quantum system $\hat{D}$ with dimensions $(2 \otimes d)$, $(\mathcal{U}_C)$ is expressed as:

$$\mathcal{U}_C(\hat{D}) = \begin{cases} 1 - \Lambda_{min}(\mathcal{W}), & \boldsymbol{r} = \boldsymbol{0} \\ 1 - \frac{1}{|\boldsymbol{r}|^2}\boldsymbol{r}\,\mathcal{W}\,\boldsymbol{r}^T, & \boldsymbol{r} \neq \boldsymbol{0} \end{cases} \tag{3}$$

Here, $\boldsymbol{r}$ represents the Bloch vector, $\boldsymbol{r}^T$ is its transpose, and $\Lambda_{\min}(\mathcal{W})$ signifies the smallest eigenvalue arising from $\mathcal{W}_{3\times 3}$ (Eq. (2)).

2.3 $\mathcal{L}_\mathcal{N}$

$\mathcal{L}_\mathcal{N}$, which is essential for quantifying entanglement, is calculated as follows:

$$\mathcal{L}_\mathcal{N}(\hat{D}) = \log_2\left(\sum_i |\nu_i|\right) \tag{4}$$

Here, ν_i represents the eigenvalues of $(\hat{D}^{T_1})$.

3 The Heisenberg Qubit-Qutrit XXZ Model with Intrinsic Decoherence (ID)

3.1 The Heisenberg Qubit-Qutrit XXZ Model

This section analyzes a qubit-qutrit Heisenberg XXZ system with magnetic fields and Dzyaloshinsky-Moriya (DM) interaction, described by the Hamiltonian

$$H_{XXZ} = J\left(s_{1x}S_{2x} + s_{1y}S_{2y} + \delta\, s_{1z}S_{2z}\right) + D_z\left(s_{1x}S_{2y} - s_{1y}S_{2x}\right) \\ + (B+b)S_{2z} + (B-b)s_{1z} \qquad (5)$$

$S_{2\mu}$ and $s_{1\mu}$ ($\mu = x; y; z$) denote the spin operators corresponding to the spin-1 and spin-$1/2$ particles, respectively. δ denotes the XXZ exchange anisotropy, D_z represents the intensity of the DM interaction along the z-axis, and J signifies the Heisenberg exchange interaction. Moreover, a static external magnetic field is indicated by the parameter B. And b represents the magnetic field's degree of inhomogeneity. We notice that we are working in units so that the parameters: b, D_z, B, δ and J are dimensionless.

3.2 Intrinsic Decoherence

Intrinsic decoherence, as proposed by Milburn, involves random unitary transformations impacting non-classical correlations. Its effects on non-classical correlations have been extensively investigated [1–4]. The evolved density matrix $\hat{D}(t)$ of the system, considering ID, is given by:

$$\hat{D}(t) = \sum_{\hat{l},\hat{m}} \exp\left[-\frac{\gamma t}{2}(\mathcal{E}_{\hat{l}} - \mathcal{E}_{\hat{m}})^2 - i(\mathcal{E}_{\hat{l}} - \mathcal{E}_{\hat{m}})t\right] \langle v_{\hat{l}}|\hat{D}(0)|v_{\hat{m}}\rangle |v_{\hat{l}}\rangle\langle v_{\hat{m}}|, \qquad (6)$$

where $\hat{D}(0)$ is the initial state characterized by the mixture parameter r. $\mathcal{E}_{\hat{l},\hat{m}}$ and $|v_{\hat{l},\hat{m}}\rangle$ represent the eigenvalues and related eigenstates of H_{XXZ}, with γ denoting the ID rate.

4 Results

In this section, we compare three measures to study quantum correlations in the system. Figures depict the effects of γ, B, and D_z on these measures. The figures illustrate that increasing rates of ID negatively impact quantum correlations. Adjusting the parameters B and D_z can alter the system's ability to withstand

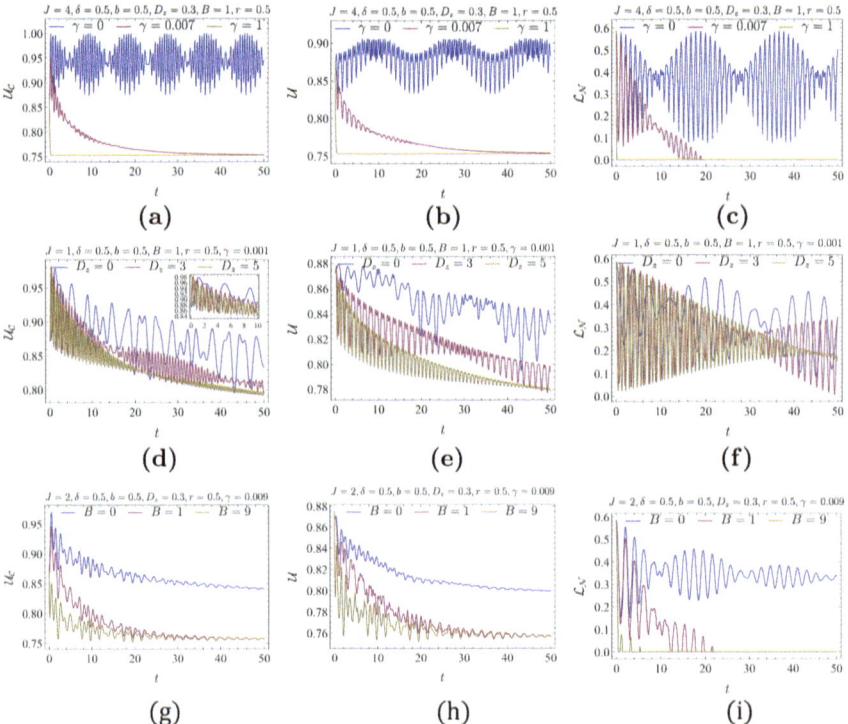

Fig. 1 The temporal evolution of $\mathcal{U}_C$, $\mathcal{U}$, and $\mathcal{L}_\mathcal{N}$ is examined for various values of γ (upper panel), D_z (middle panel), and B (lower panel)

this decoherence. Notably, reducing the values of B and D_z strengthens the system's resilience to ID, which has significant implications for quantum computing and information processing. These insights provide valuable understanding of the quantum properties in the qubit-qutrit Heisenberg XXZ model, showing promise for practical applications in quantum technology. In conclusion, our findings highlight the substantial influence of the ID parameter, magnetic field strength B, and Dzyaloshinskii-Moriya interaction magnitude D_z on both bipartite entanglement and skew information correlations within the system (Fig. 1).

References

1. E. Chaouki, M. Mansour, Skew information correlations and bipartite entanglement in hybrid qubit–qutrit system under intrinsic decoherence effect. Appl. Phys. B **129**(7), 118 (2023)
2. Z. Dahbi, M. Oumennana, M. Mansour, Intrinsic decoherence effects on correlated coherence and quantum discord in xxz heisenberg model. Opt. Quant. Electron. **55**(5), 412 (2023)
3. M. Essakhi, Y. Khedif, M. Mansour, M. Daoud, Intrinsic decoherence effects on quantum correlations dynamics. Opt. Quant. Electron. **54**, 1–15 (2022)
4. M. Oumennana, E. Chaouki, M. Mansour, The intrinsic decoherence effects on nonclassical correlations in a dipole-dipole two-spin system with dzyaloshinsky-moriya interaction. Int. J. Theor. Phys. **62**(1), 10 (2022)

Open Access This chapter is licensed under the terms of the Creative Commons Attribution 4.0 International License (http://creativecommons.org/licenses/by/4.0/), which permits use, sharing, adaptation, distribution and reproduction in any medium or format, as long as you give appropriate credit to the original author(s) and the source, provide a link to the Creative Commons license and indicate if changes were made.

The images or other third party material in this chapter are included in the chapter's Creative Commons license, unless indicated otherwise in a credit line to the material. If material is not included in the chapter's Creative Commons license and your intended use is not permitted by statutory regulation or exceeds the permitted use, you will need to obtain permission directly from the copyright holder.

Elasic Neutrino-Nucleon Scattering: The Effects of Electromagnetic Properties and Polarization

Konstantin Kouzakov, Fedor Lazarev, and Alexander Studenikin

1 Introduction

There are a large number of experiments investigating both neutrino oscillations and their interactions. In both cases, it is important to theoretically investigate neutrino scattering on various targets, since scattering processes is a tool both for studying fundamental interactions of neutrinos and for detecting neutrino fluxes: the processes of neutrino scattering on a nucleon or a nucleus contribute to the signals of such experiments as, for example, MiniBooNE, COHERENT and registration of supernova neutrinos in JUNO. Of great interest in this context is the search for the electromagnetic (EM) properties of neutrinos (see, for instance, the review article [1] and references therein). It should be noted that neutrino charge radii and magnetic moments are predicted to be nonzero already in the Standard Model and in its minimal extension, respectively. The nonzero magnetic moments may induce neutrino spin oscillations in a magnetic field. This effect can be significant both in the enviroments of neutron stars and supernovas and in interstellar and/or intergalactic magnetic fields. In addition, such oscillations can affect not only the flavour composition of neutrino fluxes but also the neutrino spin state. However, to our knowledge, the neutrino spin polarization has not been taken into account in the formalism of EM neutrino-nucleon scattering. Therefore, in this contribution we fill the indicated gap.

K. Kouzakov · F. Lazarev (✉) · A. Studenikin
Lomonosov Moscow State University, Moscow, Russia
e-mail: lazarev.fm@15physics.msu.ru

© The Author(s) 2026
Y. Tayalati, M. Gouighri (eds.), *The First African Conference on High Energy Physics*, Springer Proceedings in Physics 425,
https://doi.org/10.1007/978-3-031-88933-2_40

2 General Formulation

We consider the process where an ultrarelativistic neutrino with energy E_ν originates from a source and elastically scatters on a nucleon in a detector at energy-momentum transfer $q = (T, \mathbf{q})$. In the most general case the neutrino state in the detector can be mixed both in flavour/mass space and in spin space. In our work we assume that the neutrino is born in the source in flavour state ℓ and leaves the source in a mixed spin state such that its density matrix is diagonal in the helicity basis, which coincides with the chirality basis in the tiny neutrino mass limit. Considering vacuum oscillations on the source-detector distance $\mathcal{L}$, the neutrino state in the detector is $|\nu_\ell^{L,R}(\mathcal{L})\rangle = \sum_{k=1}^{3} U_{\ell k}^* e^{-i\frac{m_k^2}{2E_\nu}\mathcal{L}} |\nu_k^{L,R}\rangle$ with probabilities α_L for left-handed (L) and $\alpha_R = 1 - \alpha_L$ for right-handed (R) neutrinos. We assume the target nucleon to be free and at rest in the lab frame.

3 Cross Sections

Since the final massive state j of the neutrino is not resolved in the detector, the differential cross section typically measured in the scattering experiment is

$$\frac{d\sigma}{dT} = \frac{\sum_j |\mathcal{M}_j|^2}{32\pi E_\nu^2 m_N}, \qquad (1)$$

where m_N is the nucleon mass. In the calculation of the matrix element $\mathcal{M}_j$, we involve the neutrino EM charge (Q), magnetic (M), electric (E) and anapole (A) form factors $f_{Q,M,E,A}^{jk}$ of diagonal ($j = k$) and transition ($j \neq k$) types in the mass basis [1]. For the nucleon part, we employ the EM form factors $F_{Q,M,E,A}$ and the weak neutral-current Dirac, Pauli and axial form factors F_1, F_2 and G_A, respectively.

Taking into account the initial neutrino mixed spin state, the cross section can be presented as $\frac{d\sigma}{dT} = \alpha_L \frac{d\sigma^L}{dT} + \alpha_R \frac{d\sigma^R}{dT}$, where $\alpha_L + \alpha_R = 1$, $\frac{d\sigma^K}{dT}$ are cross sections for the initial left-handed ($K = L$) and right-handed ($K = R$) neutrino spin states, respectively. Both of them can be split into helicity-preserving (hp) and helicity-flipping (hf) components [2]. Here we present expressions only for right-handed neutrinos, and the case of left-handed neutrinos can be found in [3, 4] (for the sake of brevity, we omit the argument $q^2 = -2m_N T$ in the expressions for the form factors):

$$\frac{d\sigma_{hp}^R}{dT} = \frac{\alpha^2\pi}{m_N T^2} C^R \left[\left(F_Q + \frac{2T}{m_N} F_A\right)^2 + \left(F_Q - \frac{2T}{m_N} F_A\right)^2 \left(1 - \frac{T}{E_\nu}\right)^2 \right.$$

$$+ \left(\frac{4T^2}{m_N^2} F_A^2 - F_Q^2\right) \frac{m_N T}{E_\nu^2} + F_M^2 \frac{T}{2m_N} \left(2 + \frac{m_N T}{E_\nu^2} - \frac{2T}{E_\nu}\right) + F_E^2 \frac{T}{2m_N}$$
$$\times \left(2 - \frac{m_N T}{E_\nu^2} - \frac{2T}{E_\nu}\right) + \frac{4T^2}{m_N E_\nu} F_A F_M \left(2 - \frac{T}{E_\nu}\right) + \frac{2T^2}{E_\nu^2} F_Q F_M \Bigg], \tag{2}$$

$$\frac{d\sigma_{hf}^R}{dT} = \frac{\pi \alpha^2}{m_e^2} |\mu_\nu^R(\mathcal{L}, E_\nu)|^2 \Bigg[\left(\frac{1}{T} - \frac{1}{E_\nu}\right) F_Q^2 + \left(\frac{1}{T} - \frac{1}{E_\nu} - \frac{m_N}{2E_\nu^2}\right) \frac{T^2}{4m_N^2} F_A^2$$
$$- \frac{T}{2E_\nu^2} F_Q F_M + \frac{\left(2 - \frac{T}{E_\nu}\right)^2 - \frac{2m_N T}{E_\nu^2}}{8m_N} F_M^2 + \frac{\left(2 - \frac{T}{E_\nu}\right)^2}{8m_N} F_E^2 \Bigg], \tag{3}$$

where

$$|\mu_\nu^R(\mathcal{L}, E_\nu)|^2 = \sum_j \left| \sum_k U_{\ell k}^* e^{-i \frac{m_k^2}{2E_\nu} \mathcal{L}} 2m_e (f_M^{jk} + i f_E^{jk}) \right|^2,$$

$$C^R = \sum_j \left| \sum_k U_{\ell k}^* e^{-i \frac{m_k^2}{2E_\nu} \mathcal{L}} (f_Q^{jk} + q^2 f_A^{jk}) \right|^2.$$

It should be noted that only the EM interactions contribute to the right-handed neutrino-nucleon scattering cross section $\frac{d\sigma^R}{dT}$. Thus, depending on the effect of neutrino spin oscillations and the values of neutrino EM characteristics, the contribution from right-handed neutrinos to the signal in a detector can be at the same level or even greater than that from left-handed ones.

4 Conclusion

We have accounted for neutrino spin oscillations on a source-detector distance in the formalism of EM neutrino-nucleon scattering. Our results can be useful in the searches for EM properties of astrophysical neutrinos with the elastic neutrino-nucleon scattering, for example, such as the neutrino-proton interaction channel for registering supernova neutrinos in the JUNO detector.

Acknowledgments The work is made within the scientific program of the National Center for Physics and Mathematics (Section No. 8, Stage 2023-2025).
This work also Supported by the Russian Science Foundation (project # 22-22-00384).

References

1. C. Giunti, A. Studenikin, Neutrino electromagnetic interactions: a window to new physics. Rev. Mod. Phys. **87**(2), 531–591 (2015). https://doi.org/10.1103/RevModPhys.87.531
2. K.A. Kouzakov, A.I. Studenikin, Electromagnetic properties of massive neutrinos in low-energy elastic neutrino-electron scattering. Phys. Rev. D **96**(5), 099904 (2017). https://doi.org/10.1103/PhysRevD.96.099904
3. K. Kouzakov, F. Lazarev, A. Studenikin, Neutrino electromagnetic properties in elastic neutrino–proton scattering. Phys. Atom. Nucl. **86**(3), 257–265 (2023). https://doi.org/10.1134/S1063778823030122
4. K. Kouzakov, F. Lazarev, A. Studenikin, Elastic neutrino–nucleon scattering and electromagnetic properties of neutrinos. Moscow Univ. Phys. Bull. **78**(6), 797 (2023). https://doi.org/10.3103/S0027134923060103

Open Access This chapter is licensed under the terms of the Creative Commons Attribution 4.0 International License (http://creativecommons.org/licenses/by/4.0/), which permits use, sharing, adaptation, distribution and reproduction in any medium or format, as long as you give appropriate credit to the original author(s) and the source, provide a link to the Creative Commons license and indicate if changes were made.

The images or other third party material in this chapter are included in the chapter's Creative Commons license, unless indicated otherwise in a credit line to the material. If material is not included in the chapter's Creative Commons license and your intended use is not permitted by statutory regulation or exceeds the permitted use, you will need to obtain permission directly from the copyright holder.

Logarithmic Negativity Versus Quantum Discord in a System of Dipolar Coupled Spins Undergoing Intrinsic Decoherence

Mansoura Oumennana, Essalha Chaouki, and Mostafa Mansour

1 Quantum Estimators

In this section, we provide the explicit formulas used to compute the logarithmic negativity $\mathcal{L}_\mathcal{N}$ and the quantum discord $\mathcal{QD}$ for a bipartite state ϱ_{AB}.

Logarithmic Negativity

$$\mathcal{L}_\mathcal{N}(\varrho_{AB}) = \log_2 \left(\sum_i |v_i| \right) \qquad (1)$$

$\{v_i\}$ are the negative eigenvalues of the partial transpose density matrix $\varrho_{AB}^{T_2}$.

Quantum Discord

$$\mathcal{QD}(\varrho_{AB}) = min\{QD_1, QD_2\}, \qquad (2)$$

with

$$QD_i = H(\varrho_{11} + \varrho_{33}) + \sum_{k=1}^{4} \lambda_k Log_2(\lambda_k) + \mathcal{D}_i. \qquad (3)$$

M. Oumennana (✉) · E. Chaouki · M. Mansour
LHEPCM, Department of Physics, Faculty of Sciences Ain Chock, Hassan II University, Casablanca, Morocco

$\mathcal{D}_1 = H(\frac{1+\sqrt{[1-2(\varrho_{44}+\varrho_{33})]^2+4(|\varrho_{23}|+|\varrho_{14}|)^2}}{2})$, $\mathcal{D}_2 = -\sum_{n=1}^{4} \varrho_{nn} Log_2(\varrho_{nn}) - H(\varrho_{11} + \varrho_{33})$. With $H(\kappa) = -\kappa Log_2(\kappa) - (1-\kappa) Log_2(1-\kappa)$ being the binary Shannon entropy. λ_k are the eigenvalues of the density operator ϱ_{AB}.

2 Dipole-Dipole Two-Spin System Under ID

The examined model is governed by the Hamiltonian presented in Eq. (4) [1]:

$$\hat{H} = -\frac{1}{3}\vec{\sigma}1^T \cdot T \cdot \vec{\sigma}2 + D_z\left(\sigma1^x\sigma2^y - \sigma_1^y\sigma_2^x\right) + B_1\sigma_1^z + B_2\sigma_2^z, \quad (4)$$

where $T = \text{diag}(\Delta - 3\epsilon, \Delta + 3\epsilon, -2\Delta)$ represents a diagonal tensor characterized by ϵ and Δ, denoting the dipolar coupling parameters influencing the spatial and relative orientations of the spins. The parameters B_i and D_z correspond to the external magnetic field acting on qubit i and the z-component of the DM interaction, respectively. The initial state of the system is the Werner state.

$$\varrho_p^\psi(0) = \frac{1-p}{4} I_2 \otimes I_2 + p|\psi\rangle\langle\psi|, \quad (5)$$

where $0 \leq p \leq 1$ is the pureness parameter and $|\psi\rangle = \frac{1}{\sqrt{2}}(|01\rangle + |10\rangle)$ is a maximally entangled Bell state.

The evolved density matrix obtained using Milburn's decoherence model [2] is given by

$$\varrho(t) = \sum_{j,k} e^{\left[-\frac{\gamma t}{2}(\mathcal{E}_j - \mathcal{E}_k)^2 - i(\mathcal{E}_j - \mathcal{E}_k)t\right]} \langle\phi_j|\varrho_p^\psi(0)|\phi_k\rangle |\phi_j\rangle\langle\phi_k| \quad (6)$$

where $\mathcal{E}_{j,k}$ and $|\phi_{j,k}\rangle$ are the eigenvalues and eigenvectors of the Hamiltonian $\hat{H}$ (4), and γ is the intrinsic decoherence parameter.

3 Results and Discussion

Our findings [1, 3, 4] reveal that in the absence of ID ($\gamma = 0$), quantum entanglement and quantum discord are permanently preserved over time as we see that the damping and the sudden death of entanglement and correlations do not occur in this case (Fig. 1). On the other hand, as soon as we consider a non-zero ID rate ($\gamma \neq 0$), both $\mathcal{L}_\mathcal{N}$ and $\mathcal{QD}$ manifest damping oscillations due to the presence of ID (Fig. 1).

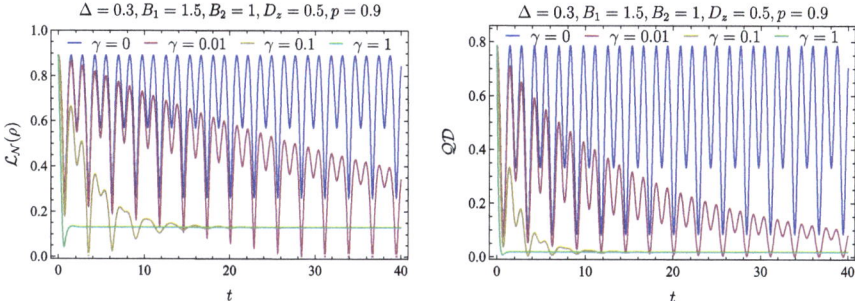

Fig. 1 $\mathcal{L}_\mathcal{N}$ (*left*) and $\mathcal{QD}$ (*right*) versus time t for different values of the ID rate γ when $\Delta = 0.3$, $B_1 = 1.5$, $B_2 = 1$, $D_z = 0.5$ and $p = 0.9$

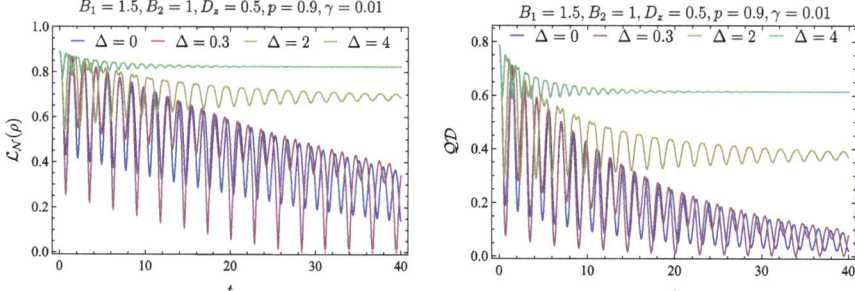

Fig. 2 $\mathcal{L}_\mathcal{N}$ (*left*) and $\mathcal{QD}$ (*right*) versus time t for different values of the dipolar parameter Δ when $\gamma = 0.01$, $B_1 = 1.5$, $B_2 = 1$, $D_z = 0.5$ and $p = 0.9$

However, increasing the dipolar parameter Δ reduces this oscillatory behavior and allows to maintain higher levels of quantum entanglement and correlations (Fig. 2). This is not the case for the DM interaction, as we notice that a strong DM interaction induces a rapid exchange of information within the dipolar-coupled spins system but does not counteract the deteriorating effect of ID on quantum correlations over time (Fig. 3). Furthermore, we find that the choice of the initial density matrix affects the quantitative dynamics of $\mathcal{L}_\mathcal{N}$ and $\mathcal{QD}$. In this case study, the choice of the Werner state $\varrho_p^\psi(0)$ (5) results in the suppression of the dipolar parameter ϵ which appears in the Hamiltonian (4) but has no effect on the evolution of $\mathcal{L}_\mathcal{N}$ and $\mathcal{QD}$. Specifically, the initial amounts recorded (at $t = 0$) of entanglement and quantum discord are exclusively affected by the pureness p of the initial state (5). Below a certain value of p, which is also contingent on the other parameters, the system is separable with zero quantum discord, or may even be separable ($\mathcal{L}_\mathcal{N}(\rho) = 0$) but still contains nonclassical correlations beyond entanglement ($\mathcal{QD}(\rho) \neq 0$) (Fig. 4).

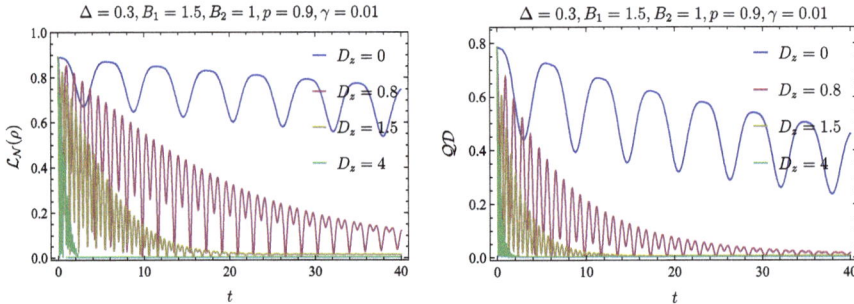

Fig. 3 $\mathcal{L_N}$ (*left*) and $\mathcal{QD}$ (*right*) versus time t for different values of the DM interaction strength D_z when $\gamma = 0.01$, $B_1 = 1.5$, $B_2 = 1$, $\Delta = 0.3$ and $p = 0.9$

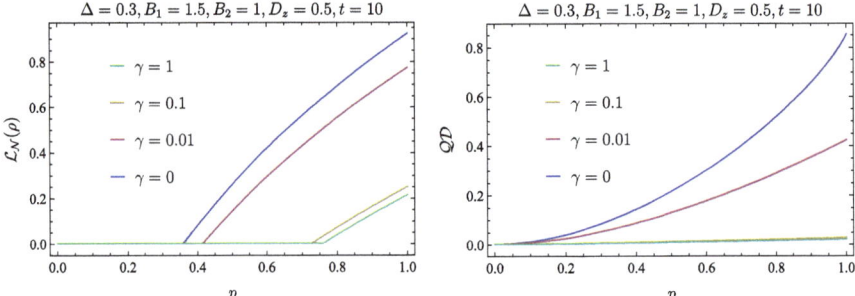

Fig. 4 $\mathcal{L_N}$ (*left*) and $\mathcal{QD}$ (*right*) versus the degree of purity p for different values of the ID rate γ when $\Delta = 0.3$, $B_1 = 1.5$, $B_2 = 1$, $D_z = 0.5$ and $t = 10$

References

1. M. Oumennana, E. Chaouki, M. Mansour, The intrinsic decoherence effects on nonclassical correlations in a dipole-dipole two-spin system with Dzyaloshinsky-Moriya interaction. Int. J. Theor. Phys. **62**(1), 10 (2022). https://doi.org/10.1007/s10773-021-04991-0
2. G.J. Milburn, Intrinsic decoherence in quantum mechanics. Phys. Rev. A **44**(9), 5401 (1991). https://doi.org/10.1103/PhysRevA.44.5401
3. Z. Dahbi, M. Oumennana, M. Mansour, Intrinsic decoherence effects on correlated coherence and quantum discord in XXZ Heisenberg model. Opt. Quant. Electron. **55**(5), 412 (2023). https://doi.org/10.1007/s11082-023-04635-y
4. E. Chaouki, Z. Dahbi, M. Mansour, Dynamics of quantum correlations in a quantum dot system with intrinsic decoherence effects. Int. J. Mod. Phys. B **36**(22), 2250141 (2022). https://doi.org/10.1142/S0217979222501417

Open Access This chapter is licensed under the terms of the Creative Commons Attribution 4.0 International License (http://creativecommons.org/licenses/by/4.0/), which permits use, sharing, adaptation, distribution and reproduction in any medium or format, as long as you give appropriate credit to the original author(s) and the source, provide a link to the Creative Commons license and indicate if changes were made.

The images or other third party material in this chapter are included in the chapter's Creative Commons license, unless indicated otherwise in a credit line to the material. If material is not included in the chapter's Creative Commons license and your intended use is not permitted by statutory regulation or exceeds the permitted use, you will need to obtain permission directly from the copyright holder.

Coherence and Non-classical Correlations within a Graphene Layer System Subjected to Intrinsic Decoherence

Zakaria Bouafia and Mostafa Mansour

1 Introduction

Two-dimensional graphene, formed of carbon atoms in a hexagonal lattice within a single layer [1], showcases extraordinary characteristics that make it ideal for various captivating applications [2]. Its enormous future potential as a nanoelectronic device makes it an exceedingly captivating material for future-generation technologies. Furthermore, its honeycomb lattice structure presents numerous appealing magnetic phenomena, including intrinsic ferromagnetism. Graphene's extended spin relaxation time makes it an appealing choice for applications involving spin qubits, as it enables more dependable storage and processing of quantum information.

Quantum correlations and coherence are fundamental to insight into the essential features of non-classical systems. From the early days of quantum mechanics to the present, quantum resources have been the focus of investigation. In this sense, local quantum uncertainty (LQU) is an appropriate way of quantifying these correlations. On the other hand, the concepts of relative entropy and l_1-norm are valuable for exploring the quantum coherence properties of non-classical systems.

In this research, we examine the dynamic evolution of LQU and coherence in a two-dimensional graphene and how they are affected by intrinsic decoherence effect.

Z. Bouafia (✉) · M. Mansour
LHEPCM, Physics Department, Faculty of Sciences of Aïn Chock, Hassan II University, Casablanca, Morocco

2 Two-Dimensional Graphene Layer System

Within the two-dimensional graphene, the unit cell comprises two separate sublattices referred to as A and B locations (refer to Fig. 1).

Both A and B function like a pseudospin system, in which each location can adopt an ascending or descending spin. The Fermi area of a half-filled honeycomb lattice features a pair of points in the first Brillouin zone, defined as K and K'. The Hamiltonian describing this system is given by the effective-mass approximation [3, 4]

$$\hat{H} = \beta[\tau(\hat{\sigma}_x \otimes \hat{\mathbb{I}})\hat{k}_x + (\hat{\sigma}_y \otimes \hat{\tau}_z)\hat{k}_y], \tag{1}$$

where $\hat{k}_x(\hat{k}_y)$ are wavenumber operators, $\beta = \hbar v_F$ with v_F represents the Fermi velocity, and $\tau = \pm 1$ denotes the valley index. We employ the decoherence model of Milburn to account for intrinsic decoherence. By employing this model, we can obtain the explicit formula of the evolved quantum state of the graphene system

$$\rho_t = \sum_{j,k} \exp\left(-\frac{\gamma t}{2}(E_j - E_k)^2 - i(E_j - E_k)t\right) \times \langle u_j | \rho^{t=0} | u_k \rangle | u_j \rangle \langle u_k |, \tag{2}$$

where γ is the intrinsic decoherence rate, $|u_{j,k}\rangle$ are the eigenstates of $\hat{H}$ with their corresponding eigenvalues $E_{j,k}$, and $\rho^{t=0}$ is the system's state at $t = 0$. We initially prepare the graphene system in a state called extended-Werner-like (EWL), which is given by

$$\rho^{t=0} = \frac{1-p}{4}\hat{\mathbb{I}}_4 + p|\Psi\rangle\langle\Psi| \quad \text{with} \quad |\Psi\rangle = \cos(\frac{\theta}{2})|00\rangle + \sin(\frac{\theta}{2})|11\rangle, \tag{3}$$

where $\hat{\mathbb{I}}_4$ is the identity matrix, the mixing parameter p with $0 \leq p \leq 1$, and $0 \leq \theta \leq \pi$.

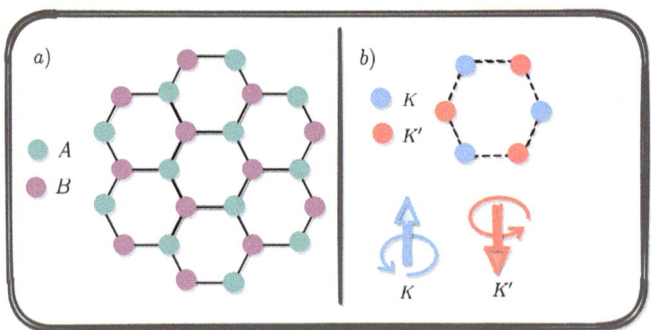

Fig. 1 (a) Structure of two-dimensional graphene made up of atoms of carbon. (b) Dirac points sites on a single cell

3 Quantum Metrics

3.1 Measures of Quantum Coherence

The l_1-norm of coherence ($C_r(\rho)$) and the relative entropy of coherence ($C_{l_1}(\rho)$) are used to assess the extent of quantum coherence in the two-dimensional graphene system. For a 2-qubits state ρ, the two measures $C_r(\rho)$ and $C_{l_1}(\rho)$ are expressed as [5]

$$C_r(\rho) = H(\rho_{diag}) - H(\rho), \tag{4}$$

$$C_{l_1}(\rho) = \sum_{i \neq j} |\rho_{i,j}|, \tag{5}$$

where $H(\rho)$ is defined by $H(\rho) = -Tr(\rho \log(\rho))$, ρ_{diag} is the diagonal state of ρ, and $\rho_{i,j}$ are the components of ρ.

3.2 LQU

To measure the quantities of non-classical correlations in the state (ρ), one uses LQU represented here by $U(\rho)$. The LQU for a given two-qubit state ρ is defined as [6, 7]

$$U(\rho) = 1 - \max(\Omega_1, \Omega_2, \Omega_3), \tag{6}$$

where $\Omega_{i=1,2,3}$ represent the eigenvalues of the matrix $\mathcal{W}$, whose elements are defined as follows

$$\left(\mathcal{W}\right)_{ij} \equiv \text{Tr}\left\{\sqrt{\rho}\left(\hat{\sigma}_{Ai} \otimes \hat{\mathbb{I}}_B\right)\sqrt{\rho}\left(\hat{\sigma}_{Aj} \otimes \hat{\mathbb{I}}_B\right)\right\}, \tag{7}$$

where $\hat{\sigma}_{Ai,j}$ denotes the Pauli matrices acting on the first subsystem A.

4 Results

We represent here the influence of various physical system parameters, including the intrinsic decoherence (γ), the parameter of purity (p), and the graphene system parameter (k_1) on the temporal evolution of $C_{l_1}(\rho)$, $C_r(\rho)$ and $U(\rho)$ in the graphene layer system. We note that $k_1 = \beta k_x$ and $k_2 = \beta k_y$.

In Fig. 2a–c, we explore how the time evolution of non-classical correlations measured by LQU ($U(\rho)$) (Fig. 2c) and quantum coherence captured by $C_{l_1}(\rho)$

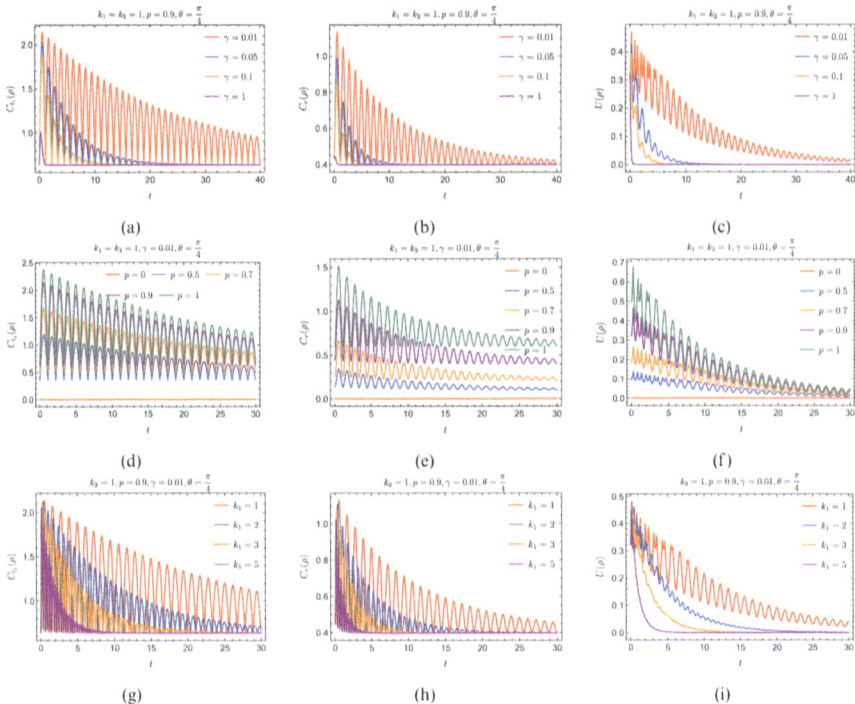

Fig. 2 Time evolution of $C_{l_1}(\rho)$, $C_r(\rho)$, and $U(\rho)$ for various values of γ (**a**)–(**c**), p (**d**)–(**f**), and k_1 (**g**)–(**i**)

(Fig. 2a) and $C_r(\rho)$ (Fig. 2b) impacted by the intrinsic decoherence rates (γ). When time (t) progresses, the three metrics present damped oscillations. We show that the amplitude of these oscillations decays with increasing the values of γ and evolves to a coherent steady state for Fig. 2a, b and towards an uncorrelated steady state for Fig. 2c. These results clearly show that intrinsic decoherence negatively affects quantum resources in the graphene system.

Next, Fig. 2d–f reveals the dynamical of $C_{l_1}(\rho)$ (Fig. 2d) and $C_r(\rho)$ (Fig. 2e) and $U(\rho)$ (Fig. 2f) versus p. From these figures, we observe that in the absence of p, the graphene state is described as separable and incoherent across time. This result highlights the crucial role of p in introducing quantum resources in our studied system. The amplitude of our quantum metric rises as p increases. When t approaches infinity, we show that $C_{l_1}(\rho)$ and $C_r(\rho)$ reach a non-zero steady value with a damped oscillation pattern, while $U(\rho)$ attain a constant value near to zero.

Finally, to delve deeper into the dynamic behavior of quantum resources concerning k_1, we have illustrated the impact of varying k_1 on $C_{l_1}(\rho)$, $C_r(\rho)$, and $U(\rho)$ in Fig. 2g–i. We can show from these figures that the oscillations in $C_{l_1}(\rho)$ (Fig. 2g) and $C_r(\rho)$ (Fig. 2h) vanish as k_1 increases and evolve towards a non-zero stable value. On the other hand, at $k_1 \geq 3$ in (Fig. 2i), $U(\rho)$ rapidly goes to zero. Consequently, diminishing k_1 produces a stronger positive effect on quantum coherence than on quantum correlations.

References

1. K.S. Novoselov, A.K. Geim, S.V. Morozov, D. Jiang, M.I. Katsnelson, I. Grigorieva, S.V. Dubonos, A.A. Firsov, Two-dimensional gas of massless Dirac fermions in graphene. Nature **438**(7065), 197–200 (2005). https://doi.org/10.1038/nature04233
2. A. Pospischil, M. Humer, M.M. Furchi, D. Bachmann, R. Guider, T. Fromherz, T. Mueller, CMOS-compatible graphene photodetector covering all optical communication bands. Nat. Photon. **7**(11), 892–896 (2013). https://doi.org/10.1038/nphoton.2013.240
3. Z. Bouafia, M. Mansour, Quantum interferometric power versus quantum correlations in a graphene layer system with a scattering process under thermal noise. Laser Phys. Lett. **20**(12), 125204 (2023). https://doi.org/10.1088/1612-202X/acfb48
4. Z. Bouafia, S. Elghaayda, M. Mansour, Effects of intrinsic decoherence on quantum coherence and correlations between spins within a two-dimensional honeycomb lattice graphene layer system. Mod. Phys. Lett. B **38**(1), 2350203 (2024). https://doi.org/10.1142/S021798492350203X
5. M. Oumennana, M. Mansour, Quantum coherence versus quantum-memory-assisted entropic uncertainty relation in a mixed spin-(1/2, 1) Heisenberg dimer. Opt. Quant. Electron. **55**(7), 594 (2023). https://doi.org/10.1007/s11082-023-04937-2
6. M. Benzahra, M. Mansour, M. Oumennana, S. Elghaayda, Quantum correlations and thermal coherence in a two-superconducting charge qubit system. Laser Phys. **33**(7), 075202 (2023). https://doi.org/10.1088/1555-6611/ace6e5
7. M. Oumennana, M. Mansour, Quantum correlations and coherence in a mixed spin-Heisenberg dimer under intrinsic decoherence. Phys. Scr. **99**(2), 025117 (2024). https://doi.org/10.1088/1402-4896/acf789

Open Access This chapter is licensed under the terms of the Creative Commons Attribution 4.0 International License (http://creativecommons.org/licenses/by/4.0/), which permits use, sharing, adaptation, distribution and reproduction in any medium or format, as long as you give appropriate credit to the original author(s) and the source, provide a link to the Creative Commons license and indicate if changes were made.

The images or other third party material in this chapter are included in the chapter's Creative Commons license, unless indicated otherwise in a credit line to the material. If material is not included in the chapter's Creative Commons license and your intended use is not permitted by statutory regulation or exceeds the permitted use, you will need to obtain permission directly from the copyright holder.

MIX
Papier aus verantwortungsvollen Quellen
Paper from responsible sources
FSC® C105338

If you have any concerns about our products,
you can contact us on
ProductSafety@springernature.com

In case Publisher is established outside the EU,
the EU authorized representative is:
**Springer Nature Customer Service Center GmbH
Europaplatz 3, 69115 Heidelberg, Germany**

Printed by Libri Plureos GmbH
in Hamburg, Germany